Agribusiness Management

Srishti Thakur
Sumit Bharat Wasnik
Priyanka Sharma
Bhavya Kush
Rebecca Nelson

Routledge
Taylor & Francis Group
LONDON AND NEW YORK

-EPH-
Elite Publishing House

First published 2025
by Routledge
2 Park Square, Milton Park, Abingdon, Oxon OX14 4RN

and by Routledge
605 Third Avenue, New York, NY 10158

Routledge is an imprint of the Taylor & Francis Group, an informa business

British Library Cataloguing-in-Publication Data
A catalogue record for this book is available from the British Library

Print edition not for sale in South Asia (India, Sri Lanka, Nepal, Bangladesh, Pakistan or Bhutan)

ISBN: 9781032789859 (hbk)
ISBN: 9781032789866 (pbk)
ISBN: 9781003490111 (ebk)

DOI: 10.4324/9781003490111

Typeset in Adobe Caslon Pro
by Elite Publishing House, Delhi

-EPH-

Contents

Preface

Welcome to the world of Agribusiness Management—a dynamic and ever-evolving field that lies at the intersection of agriculture, business, and technology. This book serves as a comprehensive guide to understanding the intricacies and challenges of managing agribusiness enterprises in the modern era.

The agricultural sector plays a pivotal role in feeding and sustaining the growing global population. With the increasing demand for food, fiber, and bioenergy, the agribusiness industry has witnessed significant transformations in recent years. Rapid technological advancements, changing consumer preferences, environmental concerns, and complex market dynamics have presented both opportunities and obstacles for agribusiness managers.

As we navigate this complex landscape, it becomes crucial for agribusiness professionals, students, and enthusiasts to develop a profound understanding of the key principles, strategies, and best practices that drive successful agribusiness operations. This book aims to equip you with the knowledge, insights, and tools necessary to excel in the field of agribusiness management.

Drawing upon the expertise of industry professionals, academic researchers, and experienced practitioners, this book covers a wide range of topics, encompassing the fundamentals of agribusiness management as well as the latest trends and emerging issues. From agricultural production and supply chain management to marketing, finance, and sustainability, each chapter provides a comprehensive exploration of its respective subject area.

Furthermore, this book delves into the unique challenges faced by agribusiness managers, such as resource management, risk mitigation, regulatory compliance, and the integration of technology. Through real-world case studies, practical examples, and thought-provoking exercises, we aim to bridge the gap between theory and practice, enabling you to apply your knowledge in diverse agribusiness contexts.

Whether you are a student seeking a career in agribusiness, an entrepreneur running an agricultural venture, or an industry professional looking to enhance your managerial skills, this book is designed to be your trusted companion. Its multidisciplinary approach ensures that it caters to the needs of individuals with varying backgrounds, blending concepts from agriculture, economics, finance, marketing, and management.

We are delighted to present this comprehensive resource to you, and we hope it serves as a valuable reference throughout your journey in agribusiness management. We encourage you to dive in, explore the chapters, engage in critical thinking, and embrace the innovative spirit that defines the agribusiness industry.

Remember, the world of agribusiness is dynamic, and success lies in the ability to adapt, innovate, and anticipate the changing needs of the global food system. We wish you a fruitful and rewarding learning experience as you embark on this exciting exploration of agribusiness management.

Editors

About the Editors

Dr. Srishti Thakur, currently working as an AP in Department of Agriculture, IIAST, Integral University, Lucknow, U.P. She completed her Ph.D. in Agribusiness Management from SHUATS, Naini, Prayagraj with SRF through ICAR. Shae has published two books, seven research papers and some abstracts in reputed journals.

Dr. Sumit Bharat Wasnik, completed his B.Sc. Horticulture from COH 2017 and M.Sc. Agriculture (Agricultural Economics) in 2019 both from Dr. PDKV, Akola, Maharashtra. He has qualified UGC-NET (Economics) and ICAR-NET (Agricultural Economics) in 2020 and 2021, respectively. He has completed his Ph.D. Agriculture (Agricultural Economics) with ICAR-SRF from IGKV, Raipur Chhattisgarh in 2022. Presently, he is working as a Guest Faculty (Agricultural Economics) in RSV College of Agriculture and Research Station, Bemetara, IGKV, Chhattisgarh.

Priyanka Sharma completed her B.Sc. in Agriculture from DAV University Jalandhar in 2018 and pursued her Agribusiness Management (ABM) from Dr. Yashwant Singh Parmar University of Horticulture and Forestry, Solan (Nauni), Himachal Pradesh in 2020. Presently, she is pursuing her PhD in Agribusiness Management from Dr Yashwant Singh Parmar University of Horticulture and Forestry, Solan(Nauni), Himachal Pradesh.

Bhavya Kush Currently pursuing Ph.D. Rural Management from Gujarat Vidyapith, Ahmedabad. To turn his career towards social cause, he pursued MBA (Rural Management) from Dr. RPCAU, Pusa, Bihar. Currently, he is also associated with Irrigation and Water Resources Department as an IEC Expert under DPMU, Atal Bhujal Yojana, Kurukshetra, Haryana which is a project on groundwater conservation by Govt of India and The World Bank.

Nelson. R, completed her B.Sc. in Forestry from SHUATS, Prayagraj, U.P. in 2015 and pursued her MBA Agribusiness Management (ABM) from SHUATS, Prayagraj, Uttar Pradesh in 2017. Presently, Currently she is pursuing Ph.D. in Agribusiness Management from Dr Y S Parmar University of Horticulture and Forestry, Solan (Nauni), H. P

1

Introduction to Agribusiness Management

Dr. Srishti Thakur, Dr. Setu Ratnam and Dr. Akanksha Singh

Assistant Professor, Department of Agriculture,
IIAST, Integral University, Lucknow

Abstract

Agribusiness management is a field that applies business principles and strategies to the agricultural sector. It encompasses various aspects such as production, marketing, finance, supply chain management, and sustainability practices. This abstract provides an overview of agribusiness management, highlighting its significance and key elements. Agribusiness management plays a vital role in optimizing agricultural operations and ensuring the economic sustainability of agricultural enterprises. It involves effectively managing production processes, utilizing resources efficiently, and adopting modern technologies to enhance productivity and reduce costs. Marketing is another crucial aspect of agribusiness management, focusing on understanding consumer demands, developing effective marketing strategies, and establishing distribution channels to maximize sales and market reach. Financial management is essential for budgeting, investment decisions, and ensuring the financial viability of agribusinesses. Supply chain management plays a critical role in coordinating activities from farm to market, ensuring efficient procurement, logistics, storage, processing, and timely delivery of agricultural products. Sustainability practices are increasingly important in agribusiness management, encompassing environmentally responsible practices, resource conservation, and compliance with regulations. The abstract emphasizes the importance of agribusiness management in supporting rural development, ensuring food security, and contributing to economic growth.

By adopting effective management strategies, agribusinesses can navigate market dynamics, comply with regulations, and promote sustainable practices.

Keywords: Agribusiness Management, Economic Sustainability, Financial Viability, Logistics Marketing Strategies.

Introduction

Agribusiness management refers to the application of business principles and strategies in the agricultural industry. It involves the planning, organizing, and controlling of agricultural enterprises to achieve profitability and sustainable growth. Agribusiness management combines the concepts of agriculture and business, integrating them to address the unique challenges and opportunities in the agricultural sector.

The primary goal of agribusiness management is to maximize the efficiency and profitability of agricultural operations while ensuring the responsible use of resources, maintaining environmental sustainability, and meeting the needs of consumers. It encompasses various aspects of the agricultural value chain, including production, processing, distribution, marketing, and sales of agricultural products and services.

Agribusiness managers are responsible for making informed decisions regarding crop and livestock production, resource allocation, financial management, market analysis, risk assessment, and strategic planning. They need to have a strong understanding of agricultural production practices, market dynamics, supply chain management, finance, and marketing strategies specific to the agricultural sector.

Key components of agribusiness management include:

» **Agricultural Production:** Agribusiness managers oversee the production process, including crop cultivation, livestock rearing, and the application of modern technologies to maximize yields and quality.

» **Supply Chain Management:** They manage the flow of agricultural products from farm to market, ensuring efficient logistics, storage, transportation, and distribution to meet consumer demand.

» **Marketing and Sales**: Agribusiness managers analyze market trends, develop marketing strategies, and establish relationships with buyers to effectively promote and sell agricultural products.

» **Financial Management:** They handle budgeting, financial analysis, and investment decisions, managing financial resources and optimizing returns on investment in agricultural enterprises.

» **Risk Assessment and Management:** Agribusiness managers evaluate potential risks such as weather conditions, price volatility, and regulatory changes, implementing strategies to minimize their impact and ensure business continuity.

» **Sustainability and Environmental Stewardship:** They consider the environmental and social implications of agricultural practices, adopting sustainable methods that protect natural resources, minimize waste, and contribute to the overall well-being of communities.

The field of agribusiness management offers diverse career opportunities, including farm management, agricultural marketing, agricultural finance, supply chain management, consulting, and research and development. Professionals in this field play a crucial role in ensuring food security, sustainable agriculture, and economic development in both local and global contexts.

Definition

The term Agribusiness Management is a combination of two words i.e. 'Agribusiness' and 'Management'

Agribusiness

Goldberg and Davis defined the term 'agribusiness as: "Agribusiness is a concept of economics which includes the sum total of all operations involved in the manufacturing and distribution of farm supplies; production operation on the farm; and the storage, processing and distribution of farm commodities and items made from them."

Agri-business involves three sectors:

1. Input sector: It deals with the supply of inputs required by the farmers for raising crops, livestock and other allied enterprises. These include seeds, fertilizers, chemicals, machinery and fuel.

2. Farm sector: It aims at producing crops, livestock and other products.

3. Product sector: It deals with various aspects like storage, processing and marketing the finished products so as to meet the dynamic needs of consumers.

Therefore, Agribusiness is sum total of all operations or activities involved in the business of production and marketing of farm supplies and farm products for achieving the targeted objectives.

To capture the full meaning of the term "agribusiness", it is important to visualise these three sectors as interrelated parts of a system in which the success of each part depends heavily on the proper functioning of the other two.

Management

There is no single definition of management. The following are some definitions of management according to various management people.

Henry Fayol who is considered as the father of principles of management, - "To manage is to forecast, to plan, to organise, to command coordinate and to control."

Frederick Winslow Taylor - "Management is knowing exactly what you want men to do and then seeing that they do it in the best and cheapest way."

Mary parker - "Management is the art of getting things done through people."

Peter Drucker - "Management is a multi-purpose organ that manages a business, manages manager and manages worker and work."

George Terry - "Management is a distinct process consisting of planning, organising actuating and controlling performance to determine and accomplish the objectives by the use of people and resources."

Therefore Agribusiness Management, applies business theories and practices to the agricultural industry to lower costs, boost profits and ensure that farm or food products are grown and distributed effectively.

Origin of Agribusiness

The origin of agribusiness management can be traced back to the development of modern agriculture and the increasing integration of business principles into the agricultural sector. While agriculture has been practiced for thousands of years, the concept of agribusiness management emerged in the mid-20th century as a response to the changing dynamics of agriculture and the need for efficient and profitable farming operations.

The term "agribusiness" was coined in the 1950s by Raymond Goldberg, an economist at Harvard University, to describe the integration of agriculture with business practices. It recognized that agriculture is not just a traditional way of life but a complex industry that involves production, processing, marketing, and distribution of agricultural products.

The growth of agribusiness management can be attributed to several factors:

Technological Advances: The advancements in agricultural technology, such as mechanization, irrigation systems, improved seed varieties, and the use of fertilizers and pesticides, increased the scale and efficiency of agricultural production. These developments required a more business-oriented approach to manage the complex processes and optimize productivity.

Market Globalization: The expansion of global markets and the increasing demand for agricultural products created opportunities for farmers to engage in international trade. Agribusiness management became crucial to meet quality standards, navigate complex supply chains, and develop marketing strategies to access domestic and international markets.

Changing Farm Structure: The consolidation and specialization of farms led to larger-scale operations and the need for professional management. As family farms gave way to corporate farming and agribusiness enterprises, the application of business principles became essential to ensure profitability and competitiveness.

Evolving Consumer Demands: Consumers began to demand higher quality, safer, and more diverse food products. Agribusiness management played a vital role in meeting these demands by implementing quality control measures, improving processing and packaging techniques, and developing marketing strategies to cater to changing consumer preferences.

Policy and Regulatory Changes: Government policies and regulations related to agriculture and food safety became more complex and stringent. Agribusiness managers needed to navigate these regulatory frameworks and ensure compliance while maintaining operational efficiency and profitability.

History of Agribusiness Management

The history of agribusiness management can be traced back to the early 20th century, as the agricultural sector underwent significant transformations and began incorporating business principles into farming practices. Here are some key milestones and developments in the history of agribusiness management:

Early Industrialization and Farm Mechanization (Late 19th to Early 20th Century): The introduction of farm machinery, such as tractors and harvesters, along with advancements in transportation and infrastructure, revolutionized agricultural practices. These developments marked the beginning of mechanized agriculture, which required management strategies to handle the increased scale and complexity of farming operations.

Emergence of Agricultural Cooperatives (Late 19th to Early 20th Century): Farmers began forming cooperative organizations to collectively market their products, access credit, and procure inputs. These cooperatives brought together farmers' resources and allowed for coordinated management efforts to enhance efficiency and bargaining power.

Scientific Agriculture and Extension Services (Late 19th to Early 20th Century): The advent of scientific research and agricultural extension services provided farmers with knowledge and expertise in crop rotation, soil fertility, pest management, and other agricultural practices. These advancements led to the adoption of more efficient farming techniques and the need for management strategies to implement and optimize their use.

Development of Agribusiness Programs (Mid-20th Century): Academic programs focused on agribusiness management began to emerge in universities, providing education and training in business principles and management practices specific to the agricultural industry. These programs aimed to equip future agribusiness managers with the knowledge and skills to navigate the challenges of running agricultural enterprises.

Rise of Multinational Corporations (Mid-20th Century): The expansion of multinational corporations into the agricultural sector brought a new dimension to agribusiness management. These corporations integrated various aspects of the agricultural value chain, including production, processing, and distribution, on a global scale. Agribusiness managers were tasked with overseeing operations across different countries and adapting to international markets and regulations.

Technological Advances and Precision Agriculture (Late 20th Century to Present): The introduction of advanced technologies, such as precision agriculture tools, satellite imagery, and data analytics, revolutionized farming practices. Agribusiness managers started using these technologies to optimize resource allocation, improve productivity, and reduce environmental impacts. The management of data and digital technologies became integral to decision-making processes.

Sustainable Agriculture and Environmental Concerns (Late 20th Century to Present): Growing environmental awareness and concerns about the sustainability of agricultural practices led to the integration of sustainable agriculture principles into agribusiness management. This involved implementing practices that conserve natural resources, reduce chemical inputs, promote biodiversity, and mitigate environmental impacts.

Scope of Agribusiness

The scope of agribusiness management is broad and encompasses various aspects of the agricultural industry. It involves applying business principles and strategies to effectively manage agricultural enterprises and address the challenges and opportunities in the sector. Here are some key areas within the scope of agribusiness management:

Agricultural Production Management: Agribusiness managers oversee the planning, organization, and control of agricultural production activities. This includes crop cultivation, livestock management, farm operations, and the implementation of modern technologies and best practices to optimize productivity, efficiency, and quality.

Supply Chain Management: Agribusiness managers handle the coordination and management of the agricultural value chain. They ensure the efficient flow of agricultural products from farm to market, including activities such as procurement, logistics, storage, processing, packaging, and distribution. Effective supply chain management is crucial to minimize waste, reduce costs, maintain product quality, and meet customer demands.

Marketing and Sales: Agribusiness managers analyze market trends, identify consumer preferences, develop marketing strategies, and establish distribution channels to promote and sell agricultural products. They undertake market research, pricing strategies, branding, advertising, and develop relationships with buyers, retailers, and wholesalers to ensure market access and maximize sales.

Financial Management: Agribusiness managers handle financial planning, budgeting, and investment decisions. They manage financial resources, assess profitability, analyze costs and returns, secure financing, and evaluate the financial viability of agricultural projects. Financial management also involves risk assessment, insurance, and hedging strategies to mitigate market and production risks.

Agribusiness Consulting and Advisory Services: Agribusiness managers provide consulting and advisory services to farmers, agricultural organizations, and agribusiness enterprises. They offer expertise in areas such as business planning, market analysis, farm management, risk assessment, technology adoption, and sustainability practices.

Policy and Government Relations: Agribusiness managers engage in policy analysis, advocacy, and government relations to influence agricultural policies and regulations. They stay informed about policy changes, monitor regulatory compliance, and participate in industry associations and initiatives to shape the business environment for agriculture.

Research and Development: Agribusiness managers contribute to research and development efforts in the agricultural sector. They collaborate with scientists, researchers, and technology providers to innovate and adopt new technologies, improve production practices, and develop sustainable agricultural solutions.

Entrepreneurship and Agribusiness Start-ups: Agribusiness management offers opportunities for entrepreneurship and the establishment of agribusiness start-ups. Managers identify market gaps, develop innovative business models, and create new ventures that provide agricultural products, services, or technology solutions.

The scope of agribusiness management is diverse and continuously evolving as the agricultural industry faces new challenges, technological advancements, and changing consumer demands. Agribusiness managers play a vital role in driving agricultural productivity, sustainability, profitability, and contributing to global food security and economic development.

Nature of Agribusiness

The nature of agribusiness management is characterized by the unique features and challenges of the agricultural industry. Here are some key aspects that define the nature of agribusiness management:

1. **Complexity and Interdependence:** Agribusiness management involves dealing with the complexity and interdependence of various factors within the agricultural value chain. It requires understanding and managing the interactions between agricultural production, supply chain logistics, marketing, finance, regulations, and environmental sustainability.

2. **Seasonality and Uncertainty:** The agricultural industry is highly influenced by seasonal variations, weather patterns, and natural events. Agribusiness managers must navigate these uncertainties, such as droughts, floods, pests, and market volatility. They need to develop strategies to mitigate risks, adapt to changing conditions, and make informed decisions under uncertain circumstances.

3. **Sustainable and Responsible Practices:** Agribusiness management is increasingly focused on sustainability and responsible resource management. Managers need to consider environmental impacts, natural resource conservation, climate change mitigation, and the well-being of local communities. Adopting sustainable practices and complying with regulations are essential aspects of agribusiness management.

4. **Technology and Innovation:** The adoption of technology and innovation is essential for agribusiness managers to enhance productivity, efficiency, and competitiveness. This includes the use of precision agriculture tools, digital platforms, data analytics, biotechnology, and automation to optimize production processes, improve decision-making, and reduce costs.

5. **Market Dynamics and Consumer Trends:** Agribusiness managers need to stay abreast of market dynamics and consumer trends. They must monitor changing consumer preferences, market demands, and emerging opportunities. Understanding market forces and adapting marketing strategies accordingly is crucial to remain competitive and meet consumer expectations.

6. **Global Perspective**: Agribusiness management often has a global perspective due to the international nature of agricultural trade. Agribusiness managers may deal with international markets, trade regulations, global supply chains, and cultural differences. They need to navigate these complexities and leverage international opportunities for market expansion and growth.

7. **Government Policies and Regulations**: Agribusiness managers must navigate government policies and regulations that govern the agricultural sector. These policies can impact production practices, food safety standards, environmental regulations, subsidies, and trade agreements. Managers need to stay informed, ensure compliance, and advocate for favourable policies that support their business objectives.

8. **Stakeholder Management:** Agribusiness managers interact with various stakeholders, including farmers, suppliers, retailers, consumers, government agencies, and community groups. Effective stakeholder management involves building relationships, addressing concerns, and collaborating to achieve common goals, such as sustainable agriculture, economic development, and food security.

The nature of agribusiness management requires a multidisciplinary approach, combining agricultural knowledge with business acumen, environmental awareness, technological literacy, and a global perspective. Agribusiness managers must navigate the complexities of the industry to ensure the profitability, sustainability, and resilience of agricultural enterprises.

Dimensions of Agribusiness Management

Agribusiness management encompasses various dimensions that are integral to the successful operation of agricultural enterprises. These dimensions represent different aspects and focus areas within the field. Here are some key dimensions of agribusiness management:

1. **Production Management:** This dimension involves the planning, organization, and control of agricultural production activities. It includes decisions related to crop selection, livestock management, farm layout, land utilization, irrigation, pest and disease control, and the adoption of modern farming techniques and technologies.

2. **Financial Management:** Financial management is crucial for the economic sustainability of agribusinesses. This dimension encompasses budgeting, cost analysis, financial planning, investment decisions, risk management, working capital management, and financial reporting. It also involves securing financing, assessing profitability, and optimizing the financial performance of agricultural enterprises.

3. **Marketing and Sales:** This dimension focuses on the strategies and activities related to the marketing and sale of agricultural products. It involves market research, understanding consumer demands and preferences, product development, branding, pricing, promotion, distribution channel management, and building relationships with buyers and retailers. Effective marketing and sales efforts are essential for achieving market penetration and maximizing revenue.

4. **Supply Chain Management:** Supply chain management is concerned with the coordination and integration of activities involved in the production, processing, storage, and distribution of agricultural products. This dimension includes procurement of inputs, logistics management, inventory control, quality assurance, packaging, transportation, and coordination with suppliers and distributors. Efficient supply chain management ensures timely delivery, minimizes costs, and maintains product quality.

5. **Human Resource Management:** Human resource management focuses on

managing and developing the workforce within agribusinesses. It includes recruitment, training, performance evaluation, employee motivation, labor relations, and creating a positive work environment. Effective human resource management ensures a skilled and engaged workforce, leading to improved productivity and organizational success.

6. **Risk Management:** Risk management involves identifying, assessing, and managing risks that can impact agricultural enterprises. This dimension includes analyzing production risks (such as weather, pests, and diseases), market risks (price volatility, demand fluctuations), financial risks, legal and regulatory risks, and environmental risks. Agribusiness managers implement risk mitigation strategies, such as insurance, hedging, diversification, and contingency planning, to minimize the impact of adverse events.

7. **Sustainable Agriculture and Environmental Stewardship:** This dimension focuses on adopting sustainable agricultural practices and managing environmental impacts. It includes resource conservation, soil and water management, biodiversity preservation, waste management, compliance with environmental regulations, and the integration of sustainable principles throughout the agribusiness value chain. Emphasizing sustainability ensures the long-term viability of agricultural enterprises and contributes to environmental stewardship.

These dimensions of agribusiness management are interconnected and interdependent. Successful agribusiness managers need to consider and balance these dimensions to achieve profitability, sustainability, and competitiveness in the agricultural industry.

Importance of Agribusiness Management

Agribusiness management plays a crucial role in the agricultural industry for several reasons. Here are some key reasons highlighting the importance of agribusiness management:

1. **Enhancing Productivity and Efficiency**: Agribusiness management focuses on optimizing agricultural production processes, utilizing resources efficiently, and adopting modern technologies and best practices. By implementing effective management strategies, agribusiness managers can enhance productivity, reduce waste, minimize costs, and improve overall efficiency in agricultural operations.

2. **Ensuring Profitability and Economic Sustainability:** Agribusiness management

helps agricultural enterprises achieve profitability and economic sustainability. Through financial planning, budgeting, cost analysis, and investment decisions, managers can ensure that resources are allocated effectively and revenues exceed costs. Profitable agribusinesses contribute to the livelihoods of farmers, rural communities, and the overall economic development of a region.

3. **Navigating Market Dynamics and Consumer Demands:** Agribusiness managers monitor market trends, analyze consumer preferences, and develop marketing strategies to meet changing demands. They identify market opportunities, develop branding and promotion strategies, establish distribution channels, and build relationships with buyers and retailers. Effective market management allows agricultural enterprises to adapt to consumer needs, access new markets, and achieve competitive advantages.

4. **Managing Supply Chain and Value Chain Coordination:** Agribusiness management encompasses supply chain management and coordination across the agricultural value chain. Managers ensure efficient procurement of inputs, effective logistics, quality control, storage, processing, and timely delivery of agricultural products. By optimizing supply chain activities, managers minimize post-harvest losses, maintain product quality, and maximize value creation throughout the value chain.

5. **Adapting to Changing Regulatory Environment:** Agribusiness managers stay updated with agricultural policies, regulations, and compliance requirements. They ensure that the business operations align with legal and regulatory frameworks related to food safety, environmental protection, labor laws, and trade regulations. Compliance with regulations is crucial for the reputation, risk management, and sustainability of agricultural enterprises.

6. **Promoting Sustainable Agriculture and Environmental Stewardship:** Agribusiness management plays a vital role in promoting sustainable agriculture and environmental stewardship. Managers implement sustainable practices, such as soil and water conservation, integrated pest management, biodiversity preservation, and responsible use of inputs. By adopting sustainable practices, agricultural enterprises contribute to environmental protection, resource conservation, and the long-term viability of the industry.

7. **Supporting Rural Development and Food Security:** Agribusiness management contributes to rural development by creating employment opportunities, improving infrastructure, and enhancing the economic well-being of rural communities.

It also plays a crucial role in ensuring food security by optimizing production, increasing agricultural productivity, and facilitating the efficient distribution of food products.

In summary, agribusiness management is important for enhancing productivity, ensuring profitability, navigating market dynamics, managing supply chains, complying with regulations, promoting sustainability, and supporting rural development and food security. Effective agribusiness management practices are essential for the overall success and sustainability of the agricultural industry.

Problems in Agribusiness Sector

The agribusiness sector faces several challenges and problems that can hinder its growth, sustainability, and effectiveness. Here are some common problems in the agribusiness sector:

1. **Market Volatility:** Agribusiness is vulnerable to market fluctuations, including unpredictable changes in commodity prices, supply and demand imbalances, and global trade uncertainties. Volatile markets can impact profitability and financial stability for agribusiness enterprises, particularly for small-scale farmers and producers who may lack the resources to withstand price shocks.

2. **Access to Finance:** Limited access to finance is a significant challenge for many agribusinesses, particularly small and medium-sized enterprises. Agricultural activities often require significant upfront investments, such as purchasing land, machinery, and seeds. However, financial institutions may perceive agriculture as a high-risk sector and be hesitant to provide loans or credit at favorable terms, making it difficult for agribusinesses to secure adequate financing for their operations and expansion.

3. **Infrastructure Deficiencies:** Inadequate infrastructure, such as transportation, storage, and irrigation facilities, poses challenges to agribusiness operations. Poor transportation networks can lead to high transportation costs, delays, and spoilage of perishable goods. Limited storage facilities result in post-harvest losses and lower product quality. Insufficient irrigation systems can restrict agricultural productivity, particularly in regions with water scarcity.

4. **Climate Change and Environmental Factors:** Agribusiness is highly dependent on weather patterns and natural resources, making it vulnerable to the impacts of climate change. Erratic weather conditions, prolonged

droughts, floods, and pest and disease outbreaks can negatively affect crop yields and livestock production. Environmental degradation, soil erosion, and water pollution also pose challenges to sustainable agricultural practices and the long-term viability of agribusiness.

5. **Technological Adoption**: Despite the potential benefits of technological advancements in agribusiness, the adoption of modern technologies and practices can be slow and uneven. Limited access to information and training, high upfront costs, and resistance to change can hinder the integration of technologies such as precision agriculture, digital farming tools, and data analytics. This limits the ability of agribusinesses to optimize productivity, efficiency, and sustainability.

6. **Policy and Regulatory Constraints:** Agribusiness is subject to a complex web of policies, regulations, and standards at national and international levels. Inconsistent or burdensome regulations, trade barriers, and unclear land tenure systems can impede agribusiness operations, hinder market access, and limit investment. Lack of supportive policies and inadequate enforcement of existing regulations can also hamper the development and competitiveness of agribusiness enterprises.

7. **Lack of Market Information and Access:** Limited access to market information, such as price trends, consumer preferences, and market demand, can hinder agribusiness decision-making. Small-scale farmers and producers, in particular, may lack the knowledge and resources to access broader markets and establish profitable market linkages. Inefficient value chains and limited market infrastructure can further restrict access to markets and reduce the bargaining power of agribusinesses.

8. **Labour Shortages and Skills Gap:** The agribusiness sector often faces challenges in attracting and retaining skilled labour. Younger generations are increasingly moving away from agriculture, resulting in labour shortages and an aging workforce. Moreover, there is a need for continuous training and capacity building to equip agribusiness stakeholders with modern agricultural practices, management skills, and knowledge of market trends.

Addressing these problems requires concerted efforts from various stakeholders, including governments, financial institutions, research organizations, and agribusiness enterprises themselves. Measures such as improved market information systems, access to finance, infrastructure development, supportive policies, capacity building, and sustainable farming practices can contribute to overcoming these challenges and fostering a thriving and resilient agribusiness sector.

Scenario of Agribusiness in India

The agribusiness sector in India plays a vital role in the country's economy, as agriculture is a significant contributor to employment, rural development, and food security. Here is a scenario of the agribusiness sector in India:

1. **Diverse Agricultural Production**: India has a diverse agricultural landscape, with a wide range of crops grown across different regions. Major crops include rice, wheat, pulses, cotton, sugarcane, oilseeds, fruits, and vegetables. Agribusinesses in India engage in various activities such as crop production, livestock rearing, fisheries, and horticulture, catering to both domestic and international markets.

2. **Smallholder Dominance:** The majority of farmers in India are smallholders with fragmented landholdings. Small-scale farming is prevalent, and agribusinesses often work with these farmers to procure agricultural produce, provide inputs, and support them with technical assistance. Agribusinesses play a crucial role in empowering smallholder farmers by offering access to markets, knowledge, and resources.

3. **Growing Organic and Sustainable Agriculture**: There is an increasing demand for organic and sustainably produced agricultural products in India and the global market. Agribusinesses are recognizing this trend and investing in organic farming practices, promoting sustainable agriculture, and obtaining organic certifications. This shift towards organic and sustainable agriculture presents opportunities for agribusinesses to meet consumer preferences and command premium prices.

4. **Technology Adoption:** Agribusinesses in India are embracing technology to enhance productivity, efficiency, and profitability. Digital agriculture solutions, precision farming techniques, and agricultural machinery are being adopted to improve farm management, optimize resource utilization, and reduce post-harvest losses. Agtech startups are emerging, offering innovative solutions such as farm management apps, remote sensing technologies, and online market platforms.

5. **Value Addition and Food Processing:** Agribusinesses in India are increasingly focusing on value addition and food processing activities. There is a growing

emphasis on transforming raw agricultural commodities into processed and packaged products, including dairy products, snacks, beverages, and ready-to-eat meals. Agribusiness enterprises are setting up food processing units, leveraging technology, and adopting quality standards to cater to the changing consumer preferences and urbanization trends.

6. **Government Initiatives:** The Indian government has implemented various initiatives to support the agribusiness sector. These include schemes for infrastructure development, irrigation, crop insurance, credit availability, and market linkages. Initiatives like the Pradhan Mantri Kisan Sampada Yojana (PMKSY) and the National Agricultural Market (e-NAM) aim to enhance the competitiveness of agribusinesses, improve farm incomes, and create a more integrated and efficient agricultural market ecosystem.

7. **Export Opportunities:** India is a significant player in global agricultural trade, exporting commodities such as rice, spices, fruits, and vegetables. Agribusinesses are actively engaged in export-oriented production and value addition to tap into international markets. The government has been promoting agricultural exports through various incentives and policies, enabling agribusinesses to explore new trade opportunities and expand their reach globally.

Despite these positive aspects, the agribusiness sector in India also faces challenges. These include fragmented landholdings, lack of market infrastructure, post-harvest losses, limited access to credit and insurance, and climate change impacts. However, efforts are being made to address these challenges through policy interventions, investments in infrastructure, and technology adoption.

In summary, the agribusiness sector in India is dynamic and evolving, with a focus on diverse agricultural production, technology adoption, value addition, and export opportunities. The sector's growth and success are critical for improving farmers' livelihoods, ensuring food security, and contributing to India's overall economic development.

Government Initiatives for Agribusiness in India

The Indian government has implemented several initiatives and programs to support and promote the agribusiness sector in the country. These initiatives aim to enhance agricultural productivity, improve market access, provide financial assistance, and address the various challenges faced by agribusinesses. Here are some key government initiatives for agribusiness in India:

1. **Pradhan Mantri Krishi Sinchayee Yojana (PMKSY):** This initiative focuses on enhancing water use efficiency in agriculture through various interventions such as micro-irrigation, precision irrigation, and watershed development. It aims to promote sustainable water management practices and improve the productivity and income of farmers.

2. **Pradhan Mantri Fasal Bima Yojana (PMFBY):** PMFBY is a crop insurance scheme that provides financial support to farmers in case of crop loss due to natural calamities, pests, or diseases. It aims to safeguard farmers' income and encourage them to adopt modern agricultural practices without the fear of financial risks.

3. **National Agricultural Market (e-NAM):** e-NAM is an online platform that integrates wholesale markets across the country, enabling farmers to sell their produce electronically. It promotes transparency, fair pricing, and market access for farmers, reducing intermediaries and improving their bargaining power.

4. **Rashtriya Krishi Vikas Yojana (RKVY):** RKVY is a centrally sponsored scheme that provides financial support to states for promoting agricultural development and innovation. It aims to strengthen infrastructure, enhance productivity, and promote agribusiness activities at the state level.

5. **National Food Security Mission (NFSM):** NFSM focuses on increasing the production and productivity of rice, wheat, and pulses to ensure food security. It provides financial assistance and technical support to farmers to adopt improved agricultural practices, use quality seeds, and enhance soil fertility.

6. **Pradhan Mantri Kisan Sampada Yojana (PMKSY):** PMKSY aims to modernize and strengthen the agri-food processing sector in India. It provides support for the creation of infrastructure facilities, including cold chains, food parks, processing units, and quality testing laboratories. The scheme aims to reduce post-harvest losses, enhance value addition, and promote agribusiness activities in the food processing sector.

7. **Kisan Credit Card (KCC) Scheme:** The KCC scheme provides farmers with access to timely and affordable credit for their agricultural and agribusiness activities. It enables farmers to have a flexible and convenient source of credit to meet their farming needs, including the purchase of inputs, machinery, and investment in agribusiness ventures.

8. **Soil Health Card Scheme:** This scheme aims to provide farmers with soil health

cards that contain information about the nutrient status of their soil. It helps farmers make informed decisions regarding the application of fertilizers and other soil management practices, leading to improved soil health and higher agricultural productivity.

These are some of the notable government initiatives that are aimed at promoting and supporting agribusiness in India. These programs are designed to address key challenges, provide financial support, enhance market access, and encourage sustainable agricultural practices, ultimately contributing to the growth and development of the agribusiness sector in the country.

Conclusion

In conclusion, agribusiness management is a multidimensional field that applies business principles and strategies to the agricultural industry. It encompasses various aspects, including production management, financial management, marketing and sales, supply chain management, human resource management, risk management, and sustainability practices. The importance of agribusiness management lies in its ability to enhance productivity, ensure profitability, navigate market dynamics, manage supply chains, comply with regulations, promote sustainability, and support rural development and food security. Agribusiness managers play a critical role in optimizing agricultural operations, making informed decisions, and addressing the challenges and opportunities in the agricultural sector. By employing effective management practices, agribusinesses can achieve economic sustainability, contribute to environmental stewardship, and meet the ever-changing demands of consumers and markets. Overall, agribusiness management is essential for the success, growth, and sustainability of agricultural enterprises and the overall development of the agricultural industry.

References

Barnard, C.S. and J.S. Nix (1973). *Farm Planning and Control*, Cambridge University Press.

Butterfield, K.L. (1910). 'Analysis of the Rural Problem' in *Report of the First Annual Meeting*, American Farm Management Association, Ames.

Case, H.C.M. and D.B. Williams (1957). *Fifty Years of Farm Management*, University of Illinois Press, Urbana.

Castle, E.N., M.H. Becker and A.G. Nelson (1987). *Farm Business Management*, 3rd edn, Macmillan, New York.

Dexter, K. and D. Barber (1960). *Farming for Profits*, Penguin, Harmondsworth.

Dillon, J.L. (1980). 'The Definition of Farm Management', *Journal of Agricultural Economics* 31:257-8.

Dillon, J.L. (1992). *The Farm as a Purposeful System*, Revd edn. Miscellaneous Publication No. 10, Department of Agricultural and Resource Economics, University of New England, Armidale.

Dillon, J.L. and J.R. Anderson (1990). *Analysis of Response in Crop and Livestock Production*, 3rd edn, Pergamon, Oxford.

Evelyn-White, H.G. (1914). *Hesiod: The Homeric Hymns and Homerica*, Heinemann, London.

George, C.S. (1968). *History of Management Thought*, Prentice-Hall, Englewood Cliffs.

Giles, T. and M. Stansfield (1995). *The Farmer as Manager*, 2nd edn, CAB International, Wallingford.

Hardaker, J.B., R.B.M. Huirne and J.R. Anderson (1997). *Coping with Risks in Agriculture*, CAB International, Wallingford.

Heady, E.O. and H.R. Jensen (1954). *Farm Management Economics*, Prentice-Hall, Englewood Cliffs.

Hellriegel, D. and J.W. Slocum (1986). *Management*, 4th edn, Addison-Wesley, Reading.

Hitt, M.A., R.D. Middlemist and R.L. Mathis (1989). *Management: Concepts and Effective Practice*, 3rd edn, West Publishing Co., Saint Paul.

Jensen, H.R. (1977). 'Farm Management and Production Economics, 1946-70', *A Survey of Agricultural Economics Literature*, University of Minnesota Press, Minneapolis, Vol. 1, pp. 1-89.

Kast, R.E. and J.E. Rosenzweig (eds) (1974). *Organization and Management: A Systems Approach*, 2nd edn, McGraw-Hill Kogakusha, Tokyo.

Kay, R.D. and W.M. Edwards (1994). *Farm Management*, 3rd edn, McGraw-Hill, New York.

Koontz, H. (1969). 'A Model for Analyzing the Universality and Transferability of Management', *Academy of Management Journal* 12:415-30.

Koontz, H. (1976). 'The Management Theory Jungle', in H. Koontz and C. O'Donnell (eds), *Management: A Book of Readings*, McGraw-Hill, New York, Ch. 3.

Makeham, J.P. and L.R. Malcolm (1993). *The Farming Game Now*, Cambridge University Press.

Mintzberg, H. (1983). *Structure in Fives: Designing Effective Organizations*, Prentice-Hall, Englewood Cliffs.

Rae, A.N. (1994). *Agricultural Management Economics: Activity Analysis and Decision Making*, CAB International, Wallingford.

Schultz, T.W. (1939). 'Theory of the Firm and Farm Management Research', *Journal of Farm Economics* 21: 570-86.

Taylor, F.W. (1911). *Principles of Scientific Management*, Harper & Bros., New York.

2 Supply and Demand Dynamics in Agribusiness Management

Shivani Singh[1] and Richa Khushwaha[2]

[1 and 2]Ph.D. Scholar, Department of Business Management, Dr Yashwant Singh Parmar University of Horticulture and Forestry, Nauni

ABSRACT

A market's two basic components are supply and demand. When creating and selling a product, supply defines how producers and manufacturers, large or small, react or behave in the marketplace. Understanding how factors influenced supply issues in the past will aid farm managers in understanding future supply chances. This chapter explains the dynamics of demand and supply in agricultural marketing The concept of supply refers to the decisions made by producers regarding the manufacture and selling of a commodity. A variety of factors influence supply decisions. These elements include the product's price, the number of producers, input costs, technical advancements, the price of alternative possible products, and uncontrollable factors such as weather. The elasticity of supply describes the link between the amount supplied and the price. Typically, the two most significant supply changers for farm products are technical development and weather.

Keywords: Marketplace, dynamic demand, factors, link

INTRODUCTION

Management can be defined as the articulate process by which any system systematically combines all of its resources in the organization, ranging from financial to human capital, by effectively and efficiently coordinating them in a manner that ensures that the system's cardinal objectives and goals are met (Murugan, 2017). The development of agricultural production concentration and its integration with financial, industrial, and commercial capital manifested itself in the strengthening of technological, economic, and social relationships between agriculture and industry and the financial sector of the economy, in a serious organizational restructuring of agriculture (first and foremost, strengthening the coordination of its specialized links), and the emergence of new forms regulating the agricultural sector (Gazetdinov et al., 2021).

Davis and Goldberg coined the term "agribusiness" in a 1957 paper presented at the Boston Distribution Conference as "the sum total of all operations involved in the production and distribution of food and fibre," which described three interdependent sectors in a global food system. It represents a three-part system comprised of (i) the agricultural input sector, (ii) the production sector, and (iii) the processing-manufacturing sector. (Sonker and hudson, 1989). Agribusiness is a developing field as well as an industry sector that is emerging as a potential career alternative for agribusiness students, with great prospects for both academic and industrial development. For the actual development of Indian agriculture, agribusiness management education is required for developing trained labor, establishing business opportunities, and reducing poverty through job creation, and industrial expansion. Nowadays, public, commercial, and cooperative organizations are looking for professionally qualified and trained agribusiness managers to help them flourish (Bairwa *et al.,* 2014).

The true measure of supply chain performance is how well operations are coordinated across the supply chain to produce value for consumers while enhancing the profitability of each link in the supply chain. Supply-chain management (SCM) is defined as "the management of the entire set of production, distribution, and marketing processes that provide a consumer with a desired product." (Cooper *et al.,* (1994). The demand for a specific product or service is essentially how much people are willing to pay for it at various rates. As a result, demand can be described as a relationship between price and quantity while holding all other variables constant. The demand relationship is the link between price and quantity demanded. Demand can be visually depicted as a downward-sloping curve with price on the vertical axis and quantity on the horizontal axis (Zaidi, 2012).

DEMAND AND SUPPLY

Supply and demand are one of the most fundamental economic principles, and it serves as the foundation of a market economy. Buyers want how much (quantity) of a product or service they can get. The quantity demanded is the amount of a product that people are willing to buy at a specific price; the demand relationship is the link between price and quantity demanded. The amount of supply shows how much the market can provide. The quantity supplied refers to the amount of a specific good that a producer is ready to supply in exchange for a specific price. The supply relationship is the relationship between pricing and how much of an item or service is provided to the market. Price is thus a representation of demand and supply (Jain, 2014). Supply chain management (SCM) emphasizes holistic management of the network of organizations engaged in creating and delivering a product to the customer, sometimes referred to as the supply chain (SC). SCM applies systems thinking to product provision, recognizing that several organizations are involved (Basnet and Sevring, 2016).

Management in Supply Chain

The current market is competitive and necessitates manufacturing companies that integrate their capabilities known as supply chain (Sellitto *et al.*, 2015), the demarcation of this network means managing relationships upstream and downstream, with the goal of being capable and sustainable by optimizing its supply and demand to offer a value-added product (Handayti *et al.*, 2015). The focus in managing supply chains in recent decades has been on the integration and efficiency of their total performance (Wynn et., 2016), that is, all linked activities from the transfer of raw materials to the final customer. A vision is required to establish a Supply Chain, which is primarily motivated by increased profits (Hamta *et al.*, 2015), performance optimizations, and risk reduction, all of which will result in a competitive advantage for the company throughout its supply network (Munasinghae *et al.*, 2019).

Management in Demand Chain

Demand Management in supply networks originates with the publications of Forrester and Burbidge in the 1950s and 1960s (Taylor and Fearne, 2006). In the 1990s, efficient customer reactions, supplier stock management, or continuous replenishment formed the foundation for modern Demand Management concepts in connection to supply chains (Nagashima *et al.*, 2015). Companies did not care ten years ago about the necessity of managing demand, coordinating efforts, and thereby solving the problem in Supply Chain Management (Juttner *et al.*, 2007). According to Hilletoyth and Fattila (2012), there is not much difference between the demand chain and the supply chain when it comes to the full network of organizations involved, thus it is not just about a forecasting methodology or the sort of demand or pattern that products follow but also depends on the management of managed collaboration within itself (Samuel, 2018).

Agribusiness and Supply Chain

The Supply Chain in agribusiness consists of producers, farmers, cooperatives, logistics centers, suppliers, intermediaries, processing, transportation, and distribution companies, and consumers who are interconnected throughout the network, where there are important factors to understand its complexity in order to project its demand because its parameters are not punctual and the daily demand management is low, such as product seasonality, the volume of supply, and perishability. Given the sector's complexity, it is well recognized that both producers and agricultural processors struggle to receive reliable and timely demand information. Entrepreneurship prospects in this area are heavily dependent on the network's relationships and the efficient use of its resources in order to control future demand (Buhayenko and Hertog, 2018).

The model proposed in that consists of the components of Demand Management *Fig. 2* is based on the sequence of operations of the agri-food supply from which the components of the equation model were structurally integrated with Demand Management and distribution management, as well as the dimensions of the demand-driven supply chain development model. The following five major processes are considered as a result of this integration: Forecasting demand, segmentation, sales and operations planning, warehousing, and transportation and distribution.

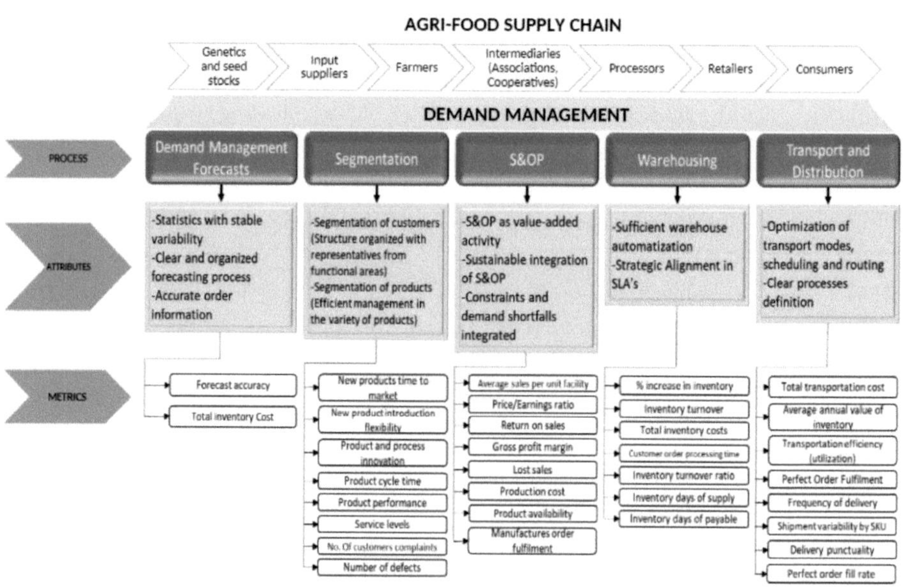

(Figure: Lank *et al.*, 2016)

Law of Demand And Supply

Law of Demand

If the Price increases, then the Quantity Demanded decreases.

If the Price decreases, then the Quantity Demanded increases.

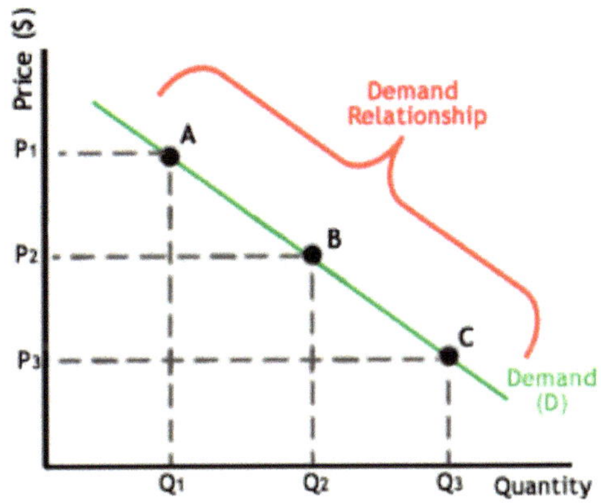

Law of Supply

If the Price increases, then the Quantity Supplied increases.

If the Price decreases, then Quantity Supplied decreases.

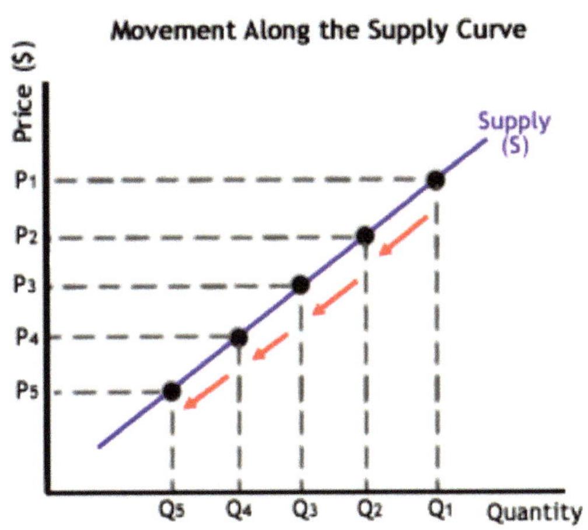

» ***Determinants of Demand***

Consumer Income

Consumer Tastes and Preference

Price of Substitute Good Price of Complementary Good

Number of Buyers

Consumer Future Expectations.

Changes in demand determinants cause the Demand Curve to alter.

Example: If consumer income rises (people have more money), demand rises (people have more money and are prepared to spend it/buy more things).

» ***Determinants of Supply***

Input Costs

Technology and Productivity

Taxes and Subsidies

Producer Future Expectations

Number of Suppliers

Changes in supply determinants will shift the Supply Curve.

Example: If Input Costs increase (it costs more to make a product), then Supply will decrease (producers have to spend more to make a product and have to cut back on production)

Relation between Demand and Supply

The link between supply and demand dictates the market price and the volume of sales of commodities or services in an unregulated market. If we take a broad view of the problem, supply and demand govern all market rivalry, thus a farmer needs to be completely knowledgeable about it.

Price Elasticity of Demand

The price elasticity of demand is the percentage change in quantity demanded caused by a 1% change in price.

$$\varepsilon = \frac{\frac{dQ}{Q}}{\frac{dp}{p}} = \frac{dQ}{dp}\frac{p}{Q},$$

Where, p denotes the price and Q(p) the market demand. Many factors can influence demand price elasticity, including the proximity of substitute items, the importance of the good in terms of expenditure, time for adjustment, product durability, and diversity of uses. In this study, we investigate the effect of reference price on price elasticity of demand, which has hitherto been overlooked. We focus on the dynamics of price elasticity as a result of changes in the quantity requested over time. Changes in demand occur when there is a price shift, and discrepancies between the new price and the consumers' reference price arise. Consumers' reference price modifications, which evolve over time, result in changes in these price deviations. The impact of these changes is reflected in the temporal dependency of quantity demanded, which causes the price elasticity dynamics. Changes in the price elasticity of demand over time have previously been explored in the marketing literature (Liu and Hanssens, 1981).

The elasticity of demand is calculated as the percentage change in quantity divided by the percentage change in the price of a product. The slope of a straight line is well recognized in the elasticity formula. However, when graphical interpretations are provided, slope and elasticity are frequently confused. When discussing third-degree price discrimination in a classroom setting, Round and McIver (2006) identify how students conflate elasticity with the slope of a demand curve. This supports the occurrence of ambiguity in distinguishing elasticity from the slope.

Economics cannot be understood without a knowledge of demand and supply theory.

It has been proposed that, when other factors stay constant, certain linkages do exist between price and amount sought and provided. But how much does the quantity given or demanded change if the price changes? It's possible that a significant price rise or decrease will have minimal impact on supply or demand for goods.

On the other hand, a minor price rise or decrease could cause a significant shift in either supply or demand. Theoretically, it is difficult to predict with certainty what will occur in situations like this. The price-quantity response may fluctuate depending on the product. It is the responsibility of economists to gather data and determine this link.

Price Elasticity of demand gauges how responsively the amount of a commodity that is demanded is to changes in price. Thus, to determine how much has changed, one must compare the percentage change in price with the consequent % change in quantity demanded.

$$n = \text{Price Elasticity} = \frac{\text{Percentage change in quantity demanded}}{\text{Percentage change in price}}$$

$$\text{or} = \frac{\text{Proportional change in quantity demanded}}{\text{Proportional change in price}}$$

$$= \frac{\Delta Q / Q}{\Delta P / P} = \frac{\Delta Q}{\Delta P} \times \frac{P}{Q} \quad (\Delta = \text{change})$$

The computation of percentage changes in this way does have a small flaw. Depending on whether we travel up or down the demand curve, we receive different results. One solution to this problem is to average two prices and two quantities over the range we are investigating, and compare the change in the average rather than the price or amount at the beginning of the change (Sanyal *et al.*, 2021).

The elasticity measure emphasizes the percentage change in quantity versus the percentage change in price. While an economic concept in elasticity highlights the responsive relationship of pricing as shown in 1(d) and 1(e), illustration 1(c) is incompatible with 1(d) and 1(e) since 1(c) depicts unit point elasticity. If the price-quantity relationship is highlighted in 1(c) with two points, the unit elasticity may not be available at a price point. The unit elasticity (Ed = 1) for a responsive connection of price with quantity can perhaps be depicted using a basic straight-line demand curve by pointing out the centre point of the demand line between the X and Y axes (Senthilnathan, 2021)

Figure 1: Elasticity of Demand in Formal Illustration

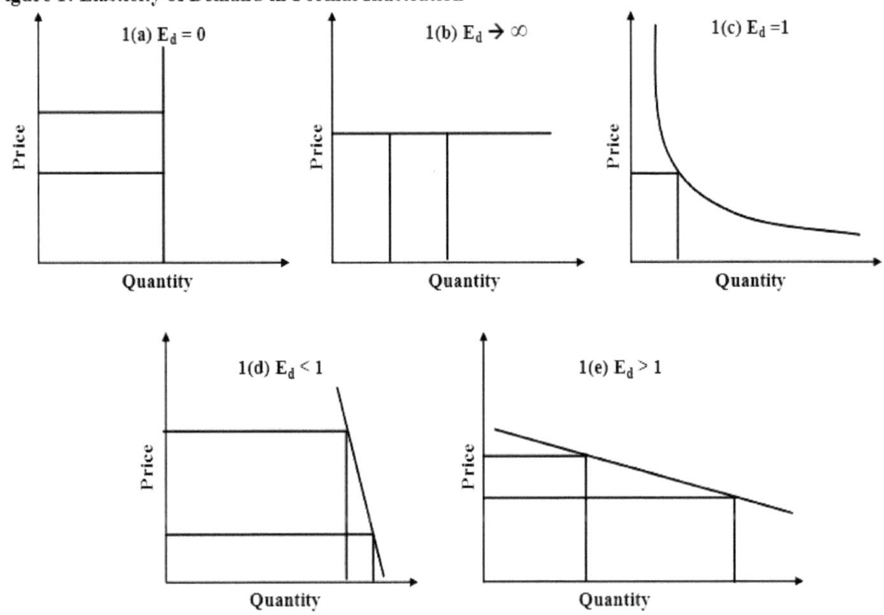

Conclusion

The quantity of goods or services that all producers are willing to offer on the market at a specific time and price is referred to as supply. Contrarily, demand is the quantity of goods or services that consumers are willing to purchase at a given price within a given time frame. Supply and demand, as well as market prices, will grow and fall until they reach a point of equilibrium, which is known as market equilibrium. For example, if orange costs are too high, most buyers will go for another fruit that is less expensive. Farmers will have to decrease their pricing in response to declining sales until demand for oranges grows again. The market is at equilibrium when the demand for oranges equals the supply. In an uncontrolled market, supply and demand determine the market price of goods or services as well as the flow of sales. As a result, this fundamental economic law has a direct impact on a farmer's profitability, and thus the success of his business.

References

Murugan MS. 2007. Management principle and practice: New York. New Age International.

Gazetdinov SM, Gazetdinov MK, Senicheva OS and Badamshin AM. 2021. Agribusiness in rural areas: Management Issues. Bio Web of Conference. 37.

Baira SL. Kalia A. Meena LK. Lakra and Kushwaha S. 2014. Agribusiness Management Education: A review on employment opportunities. International Journal of Scientific Research. 4(2):1-4.

Sanker S and Hudson MA. 1989. Why agribusiness anyway? Agribusiness. 5(4):305-314.

Cooper MC. Lamber DM and Pagh JD. 1994. Supply chain management: More than a new name for logistics. The International Journal of Logistics Management. 8(1)1-13.

Esha J. 2014. Impact of the law of demand and supply on the stock market: A study of most active BSE indices with the help of RSI. Journal of Economics and Finance. 5(6):615.

Basnet C and Seuring S. 2016. Demand-oriented supply chain strategies- A review of Literature. Operation and Supply Chain management. 9(2):73-89.

Sellito MA. Periera GM. Borchard M. Silva RI and Viega CV. 2015. ASCOR based model for supply chain performance measurement: Application in the Footwear Industry: International Journal of Production Research 53(16):4917-4926.

Handayti TM. Simatupang and Perdana T. 2015. Agrifood Supply chain Coordination: the state of the art and recent development. Logistics Research 8(1):1-15.

Wynn M and Olubanjo O. 2016. Demand and Supply Management system implication in an SME packaging business in the UK. International Journal of Manufacturing Research 5(2):198-212.

Hamta N. Shirazi MA. Ghomi SMTF and Behdad S. 2015. Supply chain network optimization considering assembly line balancing and demand uncertainty. International Journal Production Research. 53(10):2970-2994.

Munasinghae M *et al.*, 2019. Value Chain Analysis of Crude palm oil production in Brazil focusing on economic, environmental, and social. Sustainability. 17:161-176.

Taylor DH and Fearne A. 2006. Towards a framework improvement in the management of demand in agri-food supply chain. Supply chain management: An International Journal 11(5)379-384.

Nagashima M. Lassagne M Morita M and Kerbache L. 2015. Dynamic adaption of supply chain collaboration to enhance demand controllability. International Journal of Manufacturing Technology Management 29(3):139-141

Juttner O. Christopher M and Baker. 2007. Demand chain management integrating marketing and supply chain management. Industrial Marketing Management | 36(3):377-392.

Hilletofth L Latttila L. 2012. Framework for demand chain and supply coordination. International Journal of Services Sciences. 40:240-256.

Samuel KE.2018. Sharing demand signals: A new challenge to improve collaboration with the supply chain. Supply Chain Forum an International Journal 9(2):16-27.

Layenko V and Hertog DD. 2018. An adjustable robust optimization approach to optimize discount for multi-period supply chain coordination under demand uncertainty. International Journal of Production Research 56(13):458-464.

Lank K. Rao AN and Birhanu D. 2016. Supply chain performance metrics framework: A literature and research issues. International Journal of Supply Chain and Inventory Management 1(3):185.

Liu L and Hanssens D. 1981. A Bayesian approach to time-varying cross-sectional regression model. Journal of Econometrics 15:341-356.

Fibich G. Gavious A and Lowengarte O. 2005. The dynamics of price elasticity of demand in the presence of references price effect. Journal of the Academy of Marketing Science 33(1):66-78.

Senthilnatham S. 2013. An illustration of elasticity. International Journal of Innovation Research and Practises 1(3):20-31.

Sanyal RK. Mukhopadhyay and Mitrat. 2021. Demand and supply cross examination economy. International Journal of Recent Scientific Research. 12(3):4116-4117.

https://www.scribd.com/doc/88435434/Demand-and-Supply-Term-Paper-Final#

3

Human Resource Management in the Agribussiness Sector

Dr. Akanksha Singh[1], Ashish Singh[2], Ankit Pal[3] & Dr. Srishti Thakur[4]

[1]Assistant Professor, Integral Institute of Agricultural Science and Technology, Integral University, Lucknow
[2]Project Associate, CSIR-CIMAP, Lucknow
[3]Research Scholar, Department of Agricultural Extension, IARI, New Delhi
[4]Assistant Professor, Integral Institute of Agricultural Science and Technology, Integral University, Lucknow
Corresponding Author Email- ak.singh16031994@gmail.com

Abstract

"Human Resource Management is the science and the practice that deals with the nature of the employment relationship and all of the decisions, actions and issues that relate to that relationship. In practice, it involves an organization's acquisition, development and utilization of employees, as well as the employee's relationship to an organization and its performance."(Ferris R, Rosen D Sherman & Barnum T (1995).So Human Resource Management (HRM) refers to a set of programmes, functions and activities designed and carried out in order to maximize both employee as well as organizational effectiveness. "Agri-business is the sum total of all operations involved in the manufacture and distribution of farm supplies, production activities on the farm, storage, processing and distribution of farm commodities and items made from them" (John David and Gold Berg). Therefore, Agribusiness is sum total of all operations or activities involved in the business of production and marketing of farm supplies and farm products for achieving the targeted objectives. Human Resource Management (HRM) is an important aspect of any business, including agribusiness. In agribusiness, HRM involves managing and optimizing the workforce to ensure efficient and effective operations. The specific objectives of human resource management in agribusiness sector is to establish and maintain an adequate organizational structure of relationship among all the members of an organization by dividing of organization tasks into functions, positions and jobs, and by defining clearly the responsibility ,accountability, authority for each job and its relation with other jobs in the organization.

Keywords - Human Resource Management, Agribusiness, Organizations, Employees Performance, Development.

Introduction

"Human Resource Management is the science and the practice that deals with the nature of the employment relationship and all of the decisions, actions and issues that relate to that relationship. In practice, it involves an organization's acquisition, development and utilization of employees, as well as the employee's relationship to an organization and its performance."(Ferris R, Rosen D Sherman & Barnum T (1995). So Human Resource Management (HRM) refers to a set of programmes, functions and activities designed and carried out in order to maximize both employee as well as organizational effectiveness. Human Resource Management differs from personnel management both in scope and orientation. Human resource management views people as an important resource or asset to be used for the benefit of organizations, employees and society. It is emerging as a distinct philosophy of management aiming at policies that promote mutuality-mutual goals, mutual respect, mutual rewards and mutual responsibility. The belief is that policies of mutuality will elicit commitment, which in turn will yield both better economic performance and greater human resource development. Though it is a distinct philosophy, human resource cannot be treated in isolation. It is being integrated into the overall strategic management of business. Further, human resource management represents the latest term in the evolution of the subject. Personnel Management has limited scope and inverted orientation. It viewed labour as a tool, the behaviour of which could be manipulated for the benefit of the organization.

"Agri-business is the sum total of all operations involved in the manufacture and distribution of farm supplies, production activities on the farm, storage, processing and distribution of farm commodities and items made from them" (John David and Gold Berg).

Agri-business involves three sectors:

» Input sector: It deals with the supply of inputs required by the farmers for raising crops, livestock and other allied enterprises. These include seeds, fertilizers, chemicals, machinery and fuel.

» Farm sector: It aims at producing crops, livestock and other products.

» Product sector: It deals with various aspects like storage, processing and marketing the finished products so as to meet the dynamic needs of consumers.

Therefore, Agribusiness is sum total of all operations or activities involved in the business of production and marketing of farm supplies and farm products for achieving the targeted objectives.

Human Resource Management (HRM) is an important aspect of any business, including agribusiness. In agribusiness, HRM involves managing and optimizing the workforce to ensure efficient and effective operations.

Objectives of Human Resource Management for Agribusiness Sector

The basic objective of human resource management is to contribute to the reallocation of the organizational goals. However, the specific objectives of human resource management in agribusiness sector are as follows:

1. To ensure effective utilization of human resources, all other organizational resources will be efficiently utilized by the human resources.

2. To establish and maintain an adequate organizational structure of relationship among all the members of an organization by dividing of organization tasks into functions, positions and jobs, and by defining clearly the responsibility ,accountability, authority for each job and its relation with other jobs in the organization.

3. To generate maximum development of human resources within the organization by offering opportunities for advancement to employees through training and education.

4. To ensure respect for human beings by providing various services and welfare facilities to the personnel.

5. To ensure reconciliation of individual/ group goals with those of the organization in such a manner that the personnel feel a sense of commitment and loyalty towards it.

6. To identify and satisfy the needs of individuals by offering various monetary and non-monetary rewards.

In order to achieve the above objectives, human resource management under takes the following activities:

- Human Resource Planning ,i.e., determining the number and kinds of personnel required to fill various positions in the organization.
- Recruitment, selection and placement of personnel, i.e., employment function.
- Training and development of employees for their efficient performance and growth.
- Appraisal of performance of employees and taking corrective steps such as transfer from one job to another.
- Motivation of work force by providing financial incentives and avenues of promotion
- Remuneration of employees. The employees must be given sufficient wages and fringe benefits to achieve higher standard of living and to motivate them to show higher productivity.
- Social security and welfare of employees.

Need for HRM in agribusiness sector

People need competencies like knowledge, attitude, value and skill to perform task. Without continuous development of competencies in people an organization is not likely to achieve its goals. So HRM is needed by every agribusiness organization that is interested in –

- Stabilizing
- Diversifying
- Growth
- Renewing itself to become more effective
- Improving its system and services

» Change and becoming more dynamic

» Playing leadership roles

» Technological advances

» HRM is needed to develop competencies.

» To mitigate some of the evil consequences of industrialization.

» Bring about system wide change.

» To develop proper climate in the organization.

Importance of Human Resource Management in Agribusiness sector

Firms can improve organizational behaviour with effective HRM in the areas of staff commitment, flexibility and competency that may lead to potential staff performance (Koch and Mc Garth, 1996). Effective Human Resource Management practices are referred as backbone of sound HRM system. HRM practices are the organizational activities that manage the human resources in effective manner by ensuring the achievement of organizational goals (Schuler & Jackson, 1987; Schuler & MacMillan, 1984 and Wright & Snell, 1991). Khatri (1999) examined the people as factors affecting flexibility and adaptability to organizations. Managers in the organization are the adaptive mechanism that determines how the firm may respond to the competitive environment (Rundle, 1997). Managing human resource is the difficult task as compared to managing technology or capital (Barney, 1991; Lado and Wilson, 1994). Moreover, firms having competitive advantage to manage the people always acquire and deploy human resources effectively and efficiently (Wright *et al.*, 1994). Human Resource Management has a place of great importance in agribusiness sector because-

» It helps management in the preparation adoption and continuing evolution of personnel programs and policies.

» It supplies skilled workers through scientific selection process.

» It ensures maximum benefit out of the expenditure on training and development and appreciates the human assets.

» It prepares workers according to the changing needs of industry and environment.

» It motivates workers and upgrades them so as to enable them to accomplish the organization goals.

» Through innovation and experimentation in the fields of personnel, it helps in reducing casts and helps in increasing productivity.

Thus, Human Resource Management is concerned with the managing people as an organizational resource rather than as factors of production. It involves a system to be followed in business firm to recruit, select, hire, train and develop human assets. It is concerned with the people dimension of an organization. The attainment of organizational objectives depends, to agree at extent, on the way in which people are recruited, developed and utilized by the management. Therefore, proper co-ordination of human efforts and effective utilization of human and others material resources is necessary.

Some key considerations for HRM in agribusiness include:

Recruitment and Selection: HRM helps in identifying and recruiting the right talent for agribusiness. They develop job descriptions and specifications, advertise job openings, screen resumes, conduct interviews, and select the best candidates.

Training and Development: HRM provides training and development programs to employees to enhance their skills and knowledge.

Performance management: This involves setting performance goals and expectations, monitoring employee performance appraisal.

Compensation and Benefits: HR department designs methods and procedures to determine how much salary will be paid to the employees for performing specific job. To provide competitive compensation and benefits packages to attract and retain top talent.

Employee Relations: To maintain positive relationships between employer and employee. Labour relations mean interaction of managers with employees who are represented by their respective organizations. Trade unions are representatives of workers who put their efforts simultaneously to be in power for decision making for lucrative wages, benefits and other aspects of working conditions. In this regard, managers are directly responsible for negotiating with leaders of the labour unions for settling working conditions, resolving disputes and compensations issues and wages.

Career Planning: Career planning activities that are purely designed for assessment of an individual employee's potential for promotion and advancement in the firm.

With the passage of time, these activities have developed as essential source for employees to grow and excel in their fields.

Personnel Research: Human resource department is directly engaged in some sort of research activity with objective to collect facts and figures about personnel in order to develop the programmes in favour of organization. Pre-planning and post-reviewing is the main part to develop and run the personnel programme. This research is basically conducted in the areas of turnovers of employees, trainings, terminations, recruitment process etc. Opinions of the employees about rate of wages, working conditions, welfare services, job security, industrial relations, leadership etc. can be collected through surveys. These surveys help to collect real facts which assist the mangers to design more effective programmes. Although research is important, but it is one of the most neglected areas in many organizations because managers are often busy in other affairs and hence not able to spare time for this important activity. It is also said that line supervisors and executives at all levels of management also assist the human resource department for personnel research.

Compliance: Labour and employment laws and regulations are critical for compliance in all HR functions. Workplace complaints based on unfair working conditions, dissatisfaction with working conditions and unsafe employment practices are result of noncompliance.

Managing Organizational Structure: Organizational structure is the main entity that helps any organization to operate efficiently. In organization, organizational structure is the hierarchy in which people operate with various functional capacities. It also tells the values and characters of organization it believes in. Whenever, a person starts business or gets a job in the organization, it is always a great idea for him to get to know the organizational structure of the firm. Depending on the following values and nature of the business, various organizations adopt on of the following organizational structures for operational activities.

Functions of HRM in Agribusiness Sector

- Developing corporate plans, practices and cultures
- Acting as catalyst and consultant
- Acting as an expert and facilitator
- Engaging himself actively in company's strategic planning
- Keeping and sharing up-to-date information with individuals within and outside the organization

- » Identifying new corporate and HRM strategies for overall business success
- » Coordinating with different organizational teams and individuals for their effective working relationship
- » Encouraging and motivating human resources to achieve the objectives of organization effectively and efficiently
- » Diagnosing problems and issues of human resources and determining possible solutions
- » Coordinating for delivery of programmes and services for human resource development
- » Conducting, evaluating, identifying and testing how human resources have improved the organizational performance

Human resource management is all about activities to achieve the basic goals and objectives of the company as well as the needs of employees. The major role of HR manager is to integrate business practices and strategies with corporate culture and ideas that ultimately affect human resource specialists and line managers.

Role of Human Resource Management in Agribusiness

Agribusiness firms focus human resource as an expansion tool in recruitment and management of highly skilled professionals and unskilled labour. Technical and labour intensive practices are performed in agribusiness to optimize the crop production and animal husbandry from the resources invested. Therefore, each agribusiness firm requires HR departments in order to conduct proper planning of human resources required in the areas of crop and livestock production (Paul, 2014). Some of the major roles of HRM in agribusiness are;

- » **Forecasting Human Resource Needs of the Agribusiness-** The HR office works closely with the operations management team to determine the adequate number of specialized workforce and farm support laborers which are required to handle the increasing volumes of farm activities and produce marketing. To this end, HR plays the role of ensuring the appropriate personnel requirements for supporting the different technical agricultural processes and marketing activities of the agribusiness are adjusted to match the additional activities that come with expansion. HR recruits specialized employees to ensure that the skills, farm machinery and labor-intense activities are supplied with adequate number of personnel.

- **Retaining and Recruiting Personnel** - HR collaborates with the heads of sections or departments to maintain and recruit personnel but also to acquire the desired skill sets for an agribusiness workforce. HR must collaborate with the department heads because of the varied nature of the business and the skills required of the workforce. Managers of areas such as horticulture, livestock breeding, milk production, milk processing and marketing are in the best position to determine the minimum requirements and suitable remuneration packages for the specialized agricultural workers of their respective departments.
- **Implementing Policies** - HR also plays important role of ensuring that the expansion of the agribusiness is done with the support of established policies for compensating, promoting and motivating employees in different farm departments and production sections. HR ensures that the job descriptions and responsibilities of each employee are aligned to the overall production and performance targets of the business. HR collaborates with top management to ensure that employees participate fully and benefit in the growth activities, such as research and development on animal breeds, crop species and other agribusiness issues.
- **Managing Specialized Agribusiness Knowledge and Skills** -Managing knowledge involves facilitating employees to enhance their agricultural skills and work capabilities through employee mentoring programmes, research and development in new animal breeds and crop species and education advancement opportunities in agricultural fields of specialization such as agricultural engineering. The dynamic nature of agribusiness requires a personnel team capable of adjusting and living up to expectations of the ever changing business environment. HR, therefore, spearheads the efforts of supporting and keeping track of the knowledge management initiatives to ensure that personnel are continuously exposed to learning opportunities on work processes and agricultural technological developments.

Challenges of HRM in agribusiness sector

The challenges include:

- **Increasing Size of Work force:** The size of organizations is increasing. A large number of multinational organizations have grown over the years. The number of people working in the organization has also increased. The management of increased work force might create new problems and challenges as the workers are becoming more conscious of their rights.

- » **Increase in Education Level:** The governments of various countries are taking steps to eradicate illiteracy and increase the education level of their citizens. Educated consumers and workers will create very tough task for the future managers.
- » **Technological Advances:** With the changes coming in the wake of advanced technology, new jobs are created and many old jobs become redundant. In the competitive world of today, industry cannot hope to survive for long with old technology. The problem of unemployment resulting from modernization will be solved by properly assessing man power needs and training of redundant employees in alternate skills.
- » **Changes in Political Environment:** There may be greater Government‘s interference in business to safeguard the interests of workers, consumers and the public at large. Government‘s participation in trade, commerce and industry will also pose many challenges before management. The Government may restrict the scope ofprivate sector in certain areas in public interest. It does not mean chances of co-operation between the Government and private sector are ruled out.
- » **Increasing Aspirations of Employees:** Considerable changes have been noted in the worker of today in comparison to his counterpart of 1950s. The workers are becoming more aware of their higher level needs and this awareness would intensify further in the future workers.

Conclusion

In agribusiness sectors, strategic roles are performed by strategic human resource managers that increase the importance and scope of human resources management. This strategic role can only be achieved if importance of human resource is properly reorganized, restructured and reoriented along HR activities of the agribusiness firm and away from its traditionally routine activities. HR manager delineate the significant contribution of employees towards organizational success. The human resource/personnel manager must be fair and firm, tactful and resourceful, sympathetic and considerate, knowledgeable about labour laws, have broad social outlook, and have competence and confidence. Academic qualifications will also be desirable. These qualities will make him/her successful in his/her profession. In agricultural extension work, the extension personnel are the key in linking source of technology and its users. Therefore, their qualities and competencies matters most in effective transfer of technology to the clientsHRM is a growing field and complex area. Human competencies are very critical for those working in agribusiness services. Those in agribusiness service sector have to deal constantly with people. So competency development is very critical. Thus HRM need to focus on this sector for better development.

References

Aggarwal, R. A. and B. J. Simskins. (2001). Open-boon management – optimizing human capital. Business Horizons. 44: 5–13.

Barney, J. (1991). Firm resources and sustained competitive advantage. Journal of Management. 17. 99-120.

Bhide, A. and H.H. Stevenson. (1990). Why be honest if it doesn't pay. Harvard Business Review. 68:121-129.

Elaine, B. G. (2012). Five roles a leader needs to play. Article Available at www.corpmagazine.com accessed on May 10, 2023.

Fair, E. M. and L. Silvestri. (1992). Effects of Rewards, Competition and Outcome on Intrinsic Motivation. Journal of Instructional Psychology. 51:23-53.

Fuwad, J. (2014). Issues of HRM in Pakistan: New Trends in Pakistani HR Profession. Article Available at www.wordpress.com accessed on June 10, 2023.

Hackman, J.R. and G.R. Oldham. (1980). Work Redesign. Pearson Education Inc, New York, USA.

Harter, J. K., F. L. Schmidt and E. A. Killham. (2003). Employee Engagement, Satisfaction and Business-Unit-Level Outcomes: A Meta-Analysis. The Gallup Organization, Washington, USA.

Khatri, N.(1999). Emerging issues in strategic HRM in Singapore. International Journal of Manpower. 20: 516-529.

Koch, M. J. and R. G. McGrath. (1996). Improving labor productivity: human resource management policies do matter. Strategic Management Journal 17: 335–54.

Lado, A.A. and M.C. Wilson. (1994). Human resource systems and sustained competitive advantages: a competency based perspective. Academy of Management Review. 19:699-727.

Levine, D.I. (1990). Participation, productivity, and the firm's environment. California Management Review 32: 86–100.

Marks, M. L., M. L. Marks, E. J. Hackett, P. H. Mirvis and J. F. Grady. (1986). Employee participation in a quality circle program: impact on quality of work life, productivity, and absenteeism. Journal of Applied Psychology. 71: 61–69.

Osmond, V. (2013). Factors affecting organizational structure. Article Available at www.smallbusiness.chron.com accessed on May 22, 2023.

Paul, M. (2014). Role of HR in an agribusiness expansion. Article Available at www.smallbusiness.chron.com accessed on June 10, 2023.

Ramanuj, G. (2011). formal and informal forms of organization. Article Available at www.academia.edu accessed on June 3, 2023.

Rundle, S. J. (1997). Flexibility, Adaptiveness and Responsiveness (FAR-ness) as The Key Success Factors in Market Entry in The South East Asian Growth Wedge. PhD thesis, Department of Management, Monash University, Victoria.

Schuler, R. S. and S. E. Jackson. (1987). Linking competitive strategies with human resource management practices. Academy of Management Executive. 1:207- 219.

Schuler, R. S. and I. C. MacMillan. (1984). Gaining competitive advantage through human resource management practices. Human Resource Management. 23: 241- 325.

Shelley, F. (2012). The importance of training and development in the workplace. Article Available at www.smallbusiness.chron.com accessed on June 10, 2023.

Soo, J. J. R. P. (2010). The interaction effects of scheduling control and work-life balance programs on job satisfaction and mental health. International Journal of Social Welfare. 20:135–143

Traci, S. (2011). Basic types of organizational structure: formal & informal. Article Available at www.smallbusiness.chron.com accessed on June 10, 2023.

Wright, P. M., G. C. McMahan and A. McWilliams. (1994). Human resource and sustained competitive advantage: a resource based perspective. The International Journal of Human Resource Management. 5:302-26.

4

Risk Assessment and Mitigation in Agribusiness

Dr. Setu Ratnam*, Assistant Professor

Dr. Srishti Thakur* Assistant Professor

Integral Institute of Agricultural, Science and Technology IIAST, Integral University, Lucknow

Abstract

Risk assessment and mitigation play a crucial role in ensuring the success and sustainability of agribusiness ventures. Agriculture, being inherently exposed to various uncertainties and vulnerabilities, requires a systematic approach to identify, analyze, and address potential risks. This abstract provides an overview of the importance of risk assessment and mitigation in agribusiness, highlighting key steps and strategies to manage risks effectively. Overall, risk assessment and mitigation in agribusiness are essential for sustainable growth and resilience. By identifying potential risks, analyzing their implications, and implementing appropriate mitigation strategies, agribusinesses can effectively manage uncertainties, protect investments, and enhance their ability to adapt and thrive in a dynamic and volatile agricultural landscape.

Keywords: Risks in Agriculture, Risk-taking choices, risk assement, risk mitigation, government policies.

Introduction

A risk is the potential of situation or event to impact on the achievement of specific objective in the institution ot in terms of economics, low probability of expected outcome.

Types of Risks in Agriculture

Risk is an important aspect of the farming business. The uncertainties inherent in weather, yields, prices, Government policies, global markets, and other factors that impact farming can cause wide swings in farm income. Risk management involves choosing among alternatives that reduce financial effects that can result from such uncertainties.

Five general types of risk are described here: production risk, price or market risk, financial risk, institutional risk, and human or personal risk.

1. ***Production risk*** derives from the uncertain natural growth processes of crops and livestock. Here extreme weathers impact crops. Weather, disease, pests, and other factors affect both the quantity and quality of commodities produced. Hydro-meteorological calamities, including heavy rainfall and floods, have damaged 33.9 million hectares of India's cropped area between 2015-16 and 2021-22.

2. ***Price or market risk*** refers to uncertainty about the prices producers will receive for commodities or the prices they must pay for inputs. The nature of price risk varies significantly from commodity to commodity. CPI (consumer price index) based food inflation touched double digits and likewise WPI-based food inflation rose to a high of 9.3 per cent during the same period.

   a. Supply of a product is affected by a combination of production decisions made by farmers as a group and by the weather and other factors that influence yields.

   b. Demand for a product is affected by consumer preference, consumers' level of income, the strength of the general economy, and the supply and price of competing products.

   c. Cost of production of a unit of product depends on both input costs and yield. This makes it highly variable. Although input costs tend to be less variable than output prices, when combined with yield variations the cost of production becomes a serious source of risk.

3. ***Financial risk*** results when the farm business borrows money and creates an obligation to repay debt. Rising interest rates, the prospect of loans being called by lenders, and restricted credit availability are also aspects of financial risk. There is a considerable period between reward and endeavor in the

agricultural operations, particularly during the period when expenditures are generated. Production of agricultural commodities may fluctuate during this time, causing farmers' financial arrangements to be disrupted. Farmers will have to endure yet again another element of complexity as a consequence of this. Structured finance organizations use this as a pretext to reject credit for farm operations

4. ***Institutional risk*** results from uncertainties surrounding Government actions. Tax laws, regulations for chemical use, rules for animal waste disposal, and the level of price or income support payments are examples of government decisions that can have a major impact on the farm business. Part of institutional risk is the uncertainty of government policy affecting farming, such as price support and subsidies. The risks farmers face are often a result of decisions taken by policy-makers and managers. Subsidies, food quality regulations for export crops, rules for animal waste disposal and the level of price or income support payments are examples of decisions taken by government that can have a major impact on the farm business.
5. ***Human or personal risk*** refers to factors such as problems with human health or personal relationships that can affect the farm business. Accidents, illness, death, and divorce are examples of personal crises that can threaten a farm business. In many countries labour migration away from rural areas is a common occurrence. Migration can cause labour shortages for the farm. Political and social unrest can also limit labour availability.

Effective risk management in agribusiness involves a proactive approach that includes risk identification, assessment, prioritization, and the development of strategies to minimize or mitigate risks. It requires a combination of industry knowledge, data analysis, financial planning, contingency planning, and ongoing monitoring and evaluation to ensure the resilience and sustainability of agribusiness operations.

Sectors of agribusiness and their respective risks:

Sr no.	Sectors of agribusiness	Risks involved	Resultants
1	Agribusiness Input Sector (seed, fertilizer, machinery, fuel, and credit)	1. Market and Price Risks 2. Regulatory and Compliance Risks 3. Technological risks 4. Supply chain risks 5. Product quality and liability risks 6. Environmental risks 7. Financial risks	1. Fluctuating input commodity prices. 2. Regulations and compliance requirements related to product safety, labeling, environmental impact, and registration. 3. Disruption or obsolescence of existing products due to rapid innovation, intellectual property infringement (PPVFR). 4. Supply disruptions, logistical challenges, transportation delays, quality control issues, and dependence on a limited number of suppliers. 5. Product quality risks include issues such as contamination, adulteration, or substandard performance, which can lead to crop failure. 6. Concerns about pollution, runoff, and the use of potentially harmful chemicals. 7. Changes in interest rates, credit availability, exchange rates, and liquidity issues.

2	Agribusiness Output Sector (transporting, selling, storing, and inspecting/grading)	1. Market and Price Risks 2. Supply and input risks 3. Quality and safety risks 4. Regulatory and compliance risks 5. Consumer preference and trends 6. Brand reputation and public relations risks 7. Financial risks	1. Exposed to market and price risks, including fluctuations in commodity prices, changes in consumer demand, and global market conditions. 2. Risks can arise from supply disruptions, changes in input prices, fluctuations in input quality or availability, and dependence on a limited number of suppliers. 3. Risks include contamination, adulteration, foodborne illnesses, and product recalls. Failure to maintain product quality and safety can result in reputational damage, legal liabilities, and financial losses. 4. Compliance requirements related to food safety, labeling, advertising, packaging, and trade. Changes in regulations, compliance standards, and import/export regulations can impact the production processes, labeling requirements, and market access of agribusiness products. 5. Shifts in dietary habits, sustainability concerns, organic and local food movements, and ethical considerations can impact consumer choices and market demand. 6. Risks include negative publicity, product recalls, environmental controversies, and ethical concerns related to labor practices or animal welfare. 7. Sectors face financial risks, including changes in interest rates, exchange rates, credit availability, and liquidity issues. Fluctuating input costs, energy costs, and operational expenses can impact profit margins.

3	Agriservices Sector Agricultural consulting, soil testing and analysis, veterinary services, farm management services, agricultural extension and education, and agricultural finance and insurance services.	1. Business and financial risks 2. Liability risks 3. Regulatory and compliance risks 4. Technology and innovation risks 5. Weathers and climate risks 6. Reputation risks 7. Talent and workforce risks	1. Revenue fluctuations, competition, cost management, and profitability challenges. 2. Crop damage, machinery malfunctions, or advice that leads to financial loss for clients. Liability claims can result in legal disputes, financial damages, and reputational harm. 3. Failure to comply with regulations can lead to penalties, legal consequences, and loss of business. 4. Potential for technological disruptions, cybersecurity breaches, data breaches, or inadequate adoption of emerging technologies. 5. Fluctuations in weather patterns, extreme weather events, or inaccurate weather forecasts can impact service delivery and the effectiveness of recommendations. 6. Negative customer experiences, poor service quality, ethical concerns, or public controversies related to agricultural practices can damage reputation and impact client relationships 7. Shortage of skilled workers, employee turnover, and inadequate training.

Important decisions making amongst farmers:

» Acreage Decision Making

» Dynamic Decision Making Process

» Particular Impact Decisions for a Crop or Season

» Farm Planning under Risk and Uncertainty.

Farmers' attitudes toward risk

Farmers may be divided into three types: risk-neutral; risk-takers and risk-averse. The risk-averse farmers try to avoid taking risks. They tend to be more cautious individuals with preferences for less risky sources of income. In general, they will sacrifice some amount of income to reduce the chance of low income and losses. A risk averter does not refuse to accept any risk at all. However, the riskaverse farmer would seek to be compensated for the risk taken by receiving a higher return than would normally be obtained if there were no risk.

Risk-takers are people who are open to more risky business options. Unlike the risk-averse, risktakers choose the alternative that gives some chance of a higher outcome, even though they may have to accept a lower outcome. When faced with the choice, risk-taking farmers tend to prefer to take the chance to make gains rather than protecting themselves from potential losses. Even so, risk-taking farmers are still influenced by the return they could receive. Risk-neutral lies between the risk-averse and risk-taking positions. It is useful for the farmers and those who provide support services to know their attitudes towards risk. In this way, they are more conscious of the motivation behind the risk management decisions made. While most farmers tend to be risk averse, attitude concerning risk is not fixed. Many factors influence it. Thus in one situation a farmer may be risk averse, and in another situation the same farmer may be a risk-taker.

The following are some of the factors that may influence a farmer's attitude towards risk

i. Farmers who operate under subsistence conditions tend to be the most risk-averse. The provision of food for their dependants is an overriding priority for many of them. Activities with a monetary reward are frequently sacrificed in favour of meeting the objective of producing their own food.

ii. Market-oriented farmers who are not willing or able to withstand the possible financial losses associated with a risk also tend to be more risk-averse. This is often true for smallholder farmers. In effect the relationship between the input costs and the value of output from the farm influences the farmer's attitude toward risk.

iii. Family commitments and responsibilities can also play a role in attitudes toward risk. A person without family commitments may be more willing to take risks. Similarly, older people are likely to take less risks.

iv. Past experience may also influence a farmer's decisions. The effects of particularly good or bad years in the past influence decisions to be made today. Again, this may be related to age; a younger person may not yet have had many experiences on which to base decisions.

Risk-Taking Choices

In agriculture, risk-taking choices are inherent due to the uncertainties associated with various factors such as weather conditions, market fluctuations, input costs, and policy changes. Farmers and agribusinesses often make risk-taking choices to maximize their potential returns while managing and mitigating the associated risks. Here are some common risk-taking choices in agriculture:

1. **Crop Selection and Diversification**: Farmers often make risk-taking choices by selecting which crops to grow based on factors such as market demand, profitability, and their ability to manage associated risks. They may choose to diversify their crop portfolio to spread the risk across different commodities and reduce dependence on a single crop. Diversification can help offset potential losses in one crop with gains in others.

2. **Timing of Planting and Harvesting**: Decisions related to the timing of planting and harvesting crops can involve risk-taking choices. Farmers need to consider weather patterns, market conditions, and other factors that may affect yields and market prices. Planting early can carry the risk of exposure to late frosts, while delaying harvest may expose crops to adverse weather conditions.

3. **Investment in Technology and Innovation**: Agricultural producers often face risk-taking choices related to investments in new technologies and innovations. Embracing technological advancements such as precision agriculture, remote sensing, and data analytics can enhance productivity, efficiency, and risk management. However, these investments carry risks such as high upfront costs, compatibility issues, and uncertainties regarding returns on investment.

4. **Marketing and Price Risk**: Agricultural producers must make risk-taking choices when it comes to marketing their products. They need to decide when and how much to sell, considering market conditions and price fluctuations. Choices may involve forward contracting, hedging, or participating in futures markets to manage price risk. These strategies involve assumptions about future market conditions and carry their own risks.

5. **Financial Risk and Capital Investment**: Agricultural producers often make risk-taking choices when seeking financial resources for capital investments. They may take on debt or seek equity investments to fund expansion, purchase new equipment, or invest in infrastructure. These choices involve assessing financial risks such as interest rates, repayment terms, and the ability to generate sufficient returns to cover investment costs.

6. **Adoption of New Production Practices**: Farmers face risk-taking choices when deciding to adopt new production practices, such as conservation agriculture, organic farming, or alternative crop rotations. These choices involve considerations of potential benefits such as improved soil health, reduced input costs, or premium prices, as well as risks such as learning curves, production uncertainties, and market acceptance.

7. **Risk Management Strategies**: Risk-taking choices also extend to the selection and implementation of risk management strategies. Farmers and agribusinesses must decide whether to purchase crop insurance, engage in risk-sharing agreements, establish risk reserves, or use hedging instruments to manage risks associated with production, price fluctuations, or adverse weather events.

It's important to note that risk-taking choices in agriculture are influenced by individual circumstances, regional characteristics, available resources, and risk tolerance levels. Effective risk management requires a comprehensive understanding of the specific risks, informed decision-making, monitoring, and adaptability to changing conditions in order to optimize outcomes and navigate the inherent uncertainties in agriculture.

Risk Assessment and its Process

Risk assessment is a systematic process of identifying, analyzing, and evaluating potential risks associated with a specific activity, project, or operation. It involves assessing the likelihood and potential impact of various risks to determine their significance and prioritize the allocation of resources for risk management.

The primary goal of risk assessment is to identify and understand potential risks in order to make informed decisions and implement appropriate risk mitigation strategies. Here are the key steps involved in a typical risk assessment process:

i. **Risk Identification**: The first step is to identify and document potential risks relevant to the activity or project. This can be done through brainstorming sessions, reviewing historical data, conducting interviews, and engaging with stakeholders. Risks can include internal risks (e.g., operational failures, financial risks) and external risks (e.g., market changes, regulatory changes).

ii. **Risk Analysis**: Once the risks are identified, the next step is to analyze and evaluate each risk. This involves assessing the likelihood of the risk occurring and the potential impact or consequences it could have. Quantitative methods (using data and statistical analysis) and qualitative methods (expert judgment and experience) can be employed to analyze risks.

iii. **Risk Evaluation**: After analyzing the risks, they are evaluated to determine their significance and priority. This involves comparing the assessed risks against predetermined risk criteria or thresholds. Risks that exceed the acceptable thresholds are considered high-priority risks requiring immediate attention, while lower-priority risks may be monitored or addressed later.

iv. **Risk Treatment**: Risk treatment involves developing and implementing strategies to manage and mitigate identified risks. This can include risk avoidance (eliminating the risk by avoiding the activity), risk reduction (implementing measures to minimize the likelihood or impact of the risk), risk transfer (shifting the risk to another party, such as through insurance), or risk acceptance (acknowledging and monitoring the risk without taking specific mitigation actions).

v. **Risk Monitoring and Review**: Risk assessment is an ongoing process, and risks need to be monitored and reviewed periodically. This ensures that the risk assessment remains relevant and up-to-date, considering changes in the project or operational context. Monitoring and review also involve evaluating the effectiveness of risk mitigation strategies and making adjustments as necessary.

Risk assessment is a valuable tool for organizations and individuals to understand and manage potential risks effectively. By systematically identifying and analyzing risks, it helps in making informed decisions, allocating resources efficiently, and proactively addressing potential challenges to achieve desired outcomes while minimizing negative impacts.

Types of Risk Assessment in Agribusiness

In agribusiness, several types of risk assessments can be conducted to identify, analyze, and evaluate risks associated with various aspects of agricultural operations. Here are some common types of risk assessments in agribusiness:

1. **Operational Risk Assessment**: This type of risk assessment focuses on identifying risks related to day-to-day operational activities in agribusiness. It involves assessing risks associated with production processes, supply chain management, machinery and equipment, labor management, and health and safety. Operational risk assessments aim to identify potential hazards, evaluate their likelihood and consequences, and develop strategies to mitigate and manage them effectively.

2. **Market Risk Assessment**: Market risk assessments in agribusiness involve analyzing risks associated with market dynamics, including price fluctuations, demand and supply factors, competition, and changes in consumer preferences. The assessment aims to evaluate the potential impacts of market risks on the agribusiness, identify key market drivers, and develop strategies to manage price volatility, market disruptions, and changing consumer demands.

3. **Financial Risk Assessment**: Financial risk assessments focus on evaluating risks related to financial management and performance in agribusiness. This includes assessing risks associated with capital investments, financing options, credit availability, interest rates, exchange rates, and cash flow management. Financial risk assessments aim to identify potential financial vulnerabilities, evaluate risk exposure, and develop strategies to mitigate financial risks and ensure financial stability.

4. **Environmental Risk Assessment**: Environmental risk assessments in agribusiness involve evaluating risks associated with environmental factors, including soil erosion, water scarcity, pollution, and climate change impacts. The assessment aims to identify potential environmental risks, assess their potential consequences on agricultural operations, and develop strategies to minimize environmental impacts, enhance sustainability, and comply with environmental regulations.

5. **Regulatory Risk Assessment**: Regulatory risk assessments focus on identifying and evaluating risks associated with compliance with laws, regulations, and policies governing the agribusiness sector. This includes assessing risks related to food safety, labeling and packaging requirements,

pesticide and chemical use, environmental regulations, and labor laws. Regulatory risk assessments aim to ensure compliance, minimize legal and reputational risks, and develop strategies to meet regulatory obligations effectively.

6. **Technology Risk Assessment**: Technology risk assessments in agribusiness involve evaluating risks associated with the adoption and use of agricultural technologies, such as precision farming, farm management systems, and data analytics. This includes assessing risks related to data privacy and security, technological disruptions, compatibility issues, and the potential benefits and drawbacks of adopting new technologies. Technology risk assessments aim to identify potential risks, evaluate their impacts, and develop strategies to effectively manage technology-related risks.

These are some of the common types of risk assessments conducted in agribusiness. However, it's important to note that the specific types and focus of risk assessments may vary depending on the nature of the agribusiness, its size, geographical location, and specific industry sector within the agricultural value chain. Conducting regular risk assessments enables agribusinesses to proactively identify and manage risks, improve decision-making, and enhance overall operational resilience.

Methods of Risk Assessment in Agriculture

In agriculture, various methods can be employed to conduct risk assessments and evaluate the potential risks associated with agricultural operations. These methods help in identifying and analyzing risks, determining their significance, and developing appropriate risk management strategies. Here are some commonly used methods of risk assessment in agriculture:

i. **Expert Judgment**: Expert judgment involves consulting individuals with expertise and experience in specific areas of agriculture to identify and assess risks. Experts can provide insights into potential risks based on their knowledge of farming practices, regional conditions, market trends, and regulatory requirements. Expert judgment can be gathered through interviews, expert panels, or surveys to gather qualitative assessments of risks.

ii. **Historical Data Analysis**: Analyzing historical data is a quantitative method of risk assessment in agriculture. It involves studying historical records and data related to crop yields, weather patterns, market prices, input costs, and other relevant factors. By analyzing past trends and patterns, farmers and agribusinesses can gain insights into potential risks and their likelihood of occurrence.

iii. **Scenario Analysis:** Scenario analysis involves developing and evaluating different hypothetical scenarios to assess their potential impact on agricultural operations. Scenarios can be based on varying factors such as extreme weather events, market fluctuations, policy changes, or disease outbreaks. By simulating and analyzing these scenarios, potential risks and their consequences can be assessed.

iv. **Financial Modeling**: Financial modeling uses quantitative techniques to assess financial risks in agriculture. It involves projecting cash flows, estimating profitability, and conducting sensitivity analysis to evaluate the financial impacts of different risk scenarios. Financial models can help assess risks associated with input costs, market prices, interest rates, and other financial variables.

v. **On-Farm Assessments**: On-farm assessments involve conducting field visits and inspections to identify potential risks and vulnerabilities. This method involves assessing factors such as soil quality, pest and disease prevalence, crop management practices, irrigation systems, and infrastructure. By evaluating these on-farm conditions, risks related to production, yield variability, and resource management can be identified.

vi. **Participatory Approaches**: Participatory approaches involve engaging farmers, agricultural workers, and other stakeholders in the risk assessment process. This method utilizes their knowledge and experiences to identify risks and develop risk mitigation strategies. Participatory approaches can include workshops, focus group discussions, or participatory mapping exercises to gather insights and perspectives from diverse stakeholders.

vii. **Risk Checklists and Checkpoints**: Risk checklists provide a structured approach to systematically identify and evaluate potential risks. They involve using predefined lists of risk categories and specific risk factors relevant to agriculture. By going through the checklist, farmers and agribusinesses can assess their operations against potential risks and identify areas requiring further attention.

viii. **Quantitative Models and Tools**: Various quantitative models and tools can be used to assess specific risks in agriculture. Examples include crop simulation models, weather forecasting models, pest and disease prediction models, financial risk assessment tools, and decision support systems. These models and tools utilize data and algorithms to analyze and evaluate risks quantitatively.

It's important to note that different risk assessment methods can be used in combination, depending on the specific needs and context of the agricultural operation. Integrating multiple methods can provide a more comprehensive understanding of risks and improve the effectiveness of risk management strategies.

Methods of Risk Assessment in Businesses

In businesses, there are various methods of risk assessment used to identify, analyze, and evaluate risks associated with different aspects of operations. These methods help in understanding potential risks, assessing their likelihood and impact, and developing appropriate risk management strategies. Here are some commonly used methods of risk assessment in businesses:

i. **Qualitative Risk Assessment**: This method involves a subjective assessment of risks based on expert judgment and qualitative analysis. Risks are identified, described, and evaluated based on their potential impact, likelihood, and other qualitative factors. Qualitative risk assessment methods include risk matrices, risk categorization, risk scoring, and risk heat maps.

ii. **Quantitative Risk Assessment**: This method involves a numerical or quantitative analysis of risks using data and statistical techniques. It utilizes historical data, probability distributions, and mathematical models to estimate the likelihood and impact of risks. Quantitative risk assessment methods include Monte Carlo simulation, sensitivity analysis, fault tree analysis, and event tree analysis.

iii. **Delphi Technique**: The Delphi Technique involves gathering opinions and insights from a panel of experts anonymously. The experts provide their assessments and forecasts regarding potential risks, which are then compiled, analyzed, and iteratively refined to arrive at a consensus or expert consensus-based risk assessment. This method helps in aggregating expert knowledge and reducing biases.

iv. **Checklist Method:** The checklist method involves using predefined checklists or questionnaires to systematically identify and assess risks. These checklists contain a list of risk factors and questions that guide the assessment process. The method ensures that important risks are not overlooked and provides a structured approach to risk identification and evaluation.

v. **Scenario Analysis**: Scenario analysis involves developing and analyzing different hypothetical scenarios to understand their potential impact on the business. Scenarios are created based on different combinations of variables, such as market conditions, economic trends, regulatory changes, or disruptive events. By examining the consequences of these scenarios, businesses can assess risks and develop contingency plans.

vi. **Failure Mode and Effects Analysis (FMEA)**: FMEA is a systematic method used to identify and evaluate potential failures or risks in processes, products, or systems. It involves analyzing each step of a process or each component of a system to identify failure modes, assess their severity, likelihood, and detectability. FMEA helps in prioritizing risks and developing mitigation strategies to prevent or minimize failures.

vii. **Business Impact Analysis (BIA):** BIA focuses on assessing the potential impacts of risks on critical business functions, processes, and assets. It involves identifying and prioritizing critical components, estimating the potential downtime, financial losses, and reputational damage in the event of disruptions. BIA helps businesses prioritize their risk management efforts and allocate resources effectively.

viii. **External Reviews and Audits**: External reviews and audits involve engaging external experts or consultants to assess risks and evaluate the effectiveness of risk management practices. These external assessments provide an objective perspective and bring in industry best practices and benchmarks. They can identify potential risks that may be overlooked internally and provide recommendations for improvement.

It's important to note that different methods can be used in combination, depending on the specific needs, nature of the business, and industry. The selection of appropriate risk assessment methods depends on factors such as the availability of data, complexity of the risks, resources available, and the desired level of analysis and rigor.

Risk Mitigation in Agribusiness

Risk mitigation in agribusiness refers to the actions and strategies implemented to reduce or minimize the potential impact of risks on agricultural operations. It involves identifying, assessing, and prioritizing risks and then implementing measures to avoid, transfer, or control those risks. Here are some common risk mitigation strategies used in agribusiness:

1. **Diversification**: Diversification involves spreading the risk across different crops, livestock, or product lines. By diversifying the agricultural portfolio, farmers can reduce the potential impact of a single risk on their overall operations. Diversification can include growing multiple crops, raising different livestock species, or expanding into value-added products.
2. **Insurance**: Agricultural insurance provides protection against specific risks, such as crop failure, livestock mortality, or natural disasters. Farmers can purchase insurance policies that compensate them for financial losses resulting from covered risks. Insurance helps mitigate the financial impact of unforeseen events and provides a safety net for agribusinesses.
3. **Risk Transfer through Contracts:** Agribusinesses can transfer certain risks to other parties through contractual agreements. For example, forward contracts, futures contracts, or options contracts can be used to transfer price risk, ensuring a predetermined price for agricultural commodities. Supply contracts with buyers or cooperatives can also help mitigate market and demand risks.
4. **Risk Monitoring and Early Warning Systems**: Implementing risk monitoring systems and early warning mechanisms enables agribusinesses to detect potential risks at an early stage. This can involve monitoring weather patterns, market trends, pest and disease outbreaks, and other relevant indicators. Timely information and alerts allow for proactive decision-making and risk mitigation actions.
5. **Farm Management Practices**: Adopting sound farm management practices can help mitigate various risks in agribusiness. This includes implementing good agricultural practices (GAPs), proper crop rotation, soil and water conservation techniques, integrated pest management (IPM), and effective livestock management. These practices enhance the resilience of the farm and reduce the vulnerability to risks.
6. **Financial Management**: Sound financial management practices help agribusinesses mitigate financial risks. This includes maintaining appropriate financial records, budgeting, cash flow management, and strategic financial planning. By managing finances effectively, agribusinesses can withstand financial shocks and ensure stability during challenging times.
7. **Infrastructure and Equipment Maintenance**: Regular maintenance and upkeep of infrastructure, machinery, and equipment reduce the risk of breakdowns and operational disruptions. Proper maintenance practices help minimize downtime, prevent productivity losses, and ensure smooth operations.

8. **Adoption of Technology:** Embracing technological advancements and precision farming techniques can help mitigate risks and improve overall efficiency in agribusiness. This includes using remote sensing, data analytics, farm management software, and precision equipment for better decision-making, resource optimization, and risk reduction.
9. **Knowledge and Skill Development**: Enhancing knowledge and skills through training and education can help agribusinesses manage risks effectively. Staying updated with the latest industry practices, market trends, regulatory requirements, and technological advancements enables farmers and agribusiness professionals to make informed decisions and adapt to changing conditions.
10. **Collaboration and Networking**: Building networks and collaborating with other stakeholders, such as farmers' associations, cooperatives, research institutions, and government agencies, can provide access to information, resources, and support systems. Collaborative efforts can help address common risks, share best practices, and create collective risk mitigation strategies.

It's important to note that risk mitigation strategies should be tailored to the specific risks and characteristics of each agribusiness operation. Regular assessment of risks, monitoring of emerging risks, and continuous improvement of risk mitigation measures are essential for effective risk management in agribusiness.

Governmental Schemes to Mitigate Production Risks

The Department of Agriculture and Farmers Welfare, under the Government of India, has implemented several schemes to mitigate production risks and provide support to farmers. These schemes aim to enhance agricultural productivity, ensure crop and income security, and protect farmers from various risks. Here are some key schemes:

1. **Pradhan Mantri Fasal Bima Yojana (PMFBY)**: This is a crop insurance scheme aimed at providing financial protection to farmers against yield losses, crop damage, and other related risks. It covers all farmers, including sharecroppers and tenant farmers, and provides affordable insurance premiums. The scheme incorporates technology for faster claim settlements and aims to increase the coverage of insurance among farmers.

2. **Paramparagat Krishi Vikas Yojana (PKVY):** This scheme promotes organic farming practices and encourages farmers to adopt eco-friendly and sustainable

agricultural practices. It provides financial support for organic inputs, training, and capacity building. By reducing reliance on chemical inputs, the scheme helps mitigate production risks associated with pests, diseases, and soil degradation.

3. **Soil Health Card Scheme**: The Soil Health Card Scheme aims to assess the nutrient status of soils and provide customized recommendations to farmers for appropriate nutrient management. The scheme facilitates the issuance of Soil Health Cards to farmers, which provide information on the current nutrient levels, soil health, and recommendations for balanced fertilization. It helps farmers optimize fertilizer use, improve soil fertility, and mitigate production risks caused by nutrient deficiencies.

4. **Pradhan Mantri Krishi Sinchai Yojana (PMKSY):** PMKSY focuses on enhancing water use efficiency in agriculture and promoting sustainable water management practices. It provides financial assistance for the development of irrigation infrastructure, water conservation, and precision irrigation technologies. By improving access to irrigation, the scheme helps mitigate risks related to inadequate water availability and rain-dependent agriculture.

5. **National Agriculture Market (e-NAM):** e-NAM is an online platform that connects agricultural produce markets across the country, enabling transparent price discovery and efficient marketing of agricultural commodities. The platform facilitates direct access to buyers and promotes fair price realization for farmers. By reducing market risks and improving market access, e-NAM empowers farmers and enhances their income security.

6. **Rashtriya Krishi Vikas Yojana (RKVY):** RKVY supports various agricultural development initiatives at the state level, including the promotion of innovative farming practices, strengthening of agricultural infrastructure, capacity building, and market linkages. It aims to enhance productivity, diversify agriculture, and reduce risks faced by farmers through comprehensive agricultural development programs.

7. **National Agricultural Insurance Scheme (NAIS):** NAIS is an erstwhile crop insurance scheme that provides coverage against yield losses due to natural calamities, pests, and diseases. It has now been replaced by PMFBY but is still operational in some states. The scheme aims to provide affordable insurance coverage to farmers and protect them from production risks.

These are just a few examples of the schemes implemented by the Department of Agriculture and Farmers Welfare to mitigate production risks. The government continues to introduce new schemes and initiatives to support farmers and enhance agricultural resilience. Farmers can consult the respective state agricultural departments or the official website of the Ministry of Agriculture and Farmers Welfare for detailed information on schemes specific to their region.

Governmental Schemes to Mitigate the Marketing Risks

The Government of India has introduced several schemes to mitigate marketing risks in agricultural products and ensure better market access, price realization, and income security for farmers. These schemes aim to facilitate efficient marketing, reduce intermediaries, and provide market infrastructure and support. Here are some key government schemes in India focused on mitigating marketing risks in agricultural products:

1. **Agricultural Produce Market Committee (APMC) Reforms**: The government has implemented reforms to modernize agricultural markets and promote efficient marketing practices. These reforms aim to create a barrier-free market for agricultural produce, allowing farmers to sell their produce directly to buyers, including processors, exporters, and large retailers. The reforms facilitate the establishment of private markets, contract farming, and online trading platforms to enhance transparency, price discovery, and market access for farmers.

2. **Electronic National Agriculture Market (e-NAM):** e-NAM is an online platform that integrates agricultural produce markets across the country, enabling transparent price discovery and efficient marketing of agricultural commodities. It provides a common market platform for buyers and sellers, including farmers, traders, and processors. e-NAM helps mitigate marketing risks by reducing information asymmetry, enabling direct access to buyers, and promoting fair price realization for farmers.

3. **Market Intervention Scheme (MIS):** The MIS is implemented by the government to provide price support and protect farmers against price fluctuations for selected agricultural commodities. Under this scheme, when market prices fall below the specified support price, government agencies procure the produce directly from farmers at the support price. The MIS helps farmers mitigate price risks and ensures minimum price support for their produce.

4. **Price Stabilization Fund (PSF):** The PSF is established to provide support

to farmers by procuring commodities in times of excessive price volatility. The fund is utilized to intervene in the market and stabilize prices by offloading surplus stocks or creating strategic reserves. This scheme helps in mitigating risks associated with sudden price fluctuations and ensures stability in the agricultural market.

5. **Market Infrastructure Development Scheme (MIDH):** The MIDH provides financial assistance to states and union territories for the development and upgradation of agricultural marketing infrastructure. It supports the establishment of market yards, cold storages, grading and sorting facilities, market information systems, and other infrastructure components. The scheme aims to improve market infrastructure, reduce post-harvest losses, enhance value addition, and create better marketing opportunities for farmers.

6. **Operation Greens:** Operation Greens is an initiative launched to stabilize the prices of perishable horticultural commodities, such as tomatoes, onions, and potatoes (TOP crops). The scheme focuses on creating value chains from production to consumption, with an emphasis on reducing price volatility and ensuring remunerative prices for farmers. It includes interventions such as price forecasting, quality improvement, transportation, and storage infrastructure development.

7. **Pradhan Mantri Kisan Sampada Yojana (PMKSY):** PMKSY is a comprehensive scheme that aims to develop modern infrastructure for food processing, preservation, and value addition. The scheme provides financial support for setting up food processing units, cold chains, and agro-processing clusters. By promoting food processing and value addition, PMKSY facilitates better marketing opportunities, reduces post-harvest losses, and enhances farmers' income security.

These schemes and initiatives by the Government of India play a crucial role in mitigating marketing risks, improving market access, and ensuring better price realization for agricultural products. Farmers can access these schemes through the respective state agricultural departments or visit the official website of the Ministry of Agriculture and Farmers Welfare for detailed information and application procedures.

Initiatives by the Government of India to Mitigate the Financial Risks Amongst Farmers

The Government of India has implemented several efforts and schemes to mitigate financial risks among farmers and provide them with financial support and security. These initiatives aim to address challenges related to agricultural credit, debt burden, income fluctuations, and financial vulnerabilities. Here are some key efforts and schemes by the Indian government to mitigate financial risks among farmers:

i. **Kisan Credit Card (KCC) Scheme**: The KCC scheme provides farmers with easy access to short-term credit for agricultural and allied activities. It enables farmers to avail themselves of timely and adequate credit for their farming needs, including crop production, post-harvest expenses, and working capital. KCC holders benefit from interest subsidies, flexible repayment terms, and simplified documentation.

ii. **Interest Subvention Scheme (ISS):** The ISS aims to reduce the financial burden on farmers by providing interest subvention on short-term crop loans. Under this scheme, farmers receive loans at a reduced interest rate, and the government reimburses the interest subsidy to banks. It encourages timely repayment and offers interest relief to farmers during challenging periods.

iii. **Pradhan Mantri Fasal Bima Yojana (PMFBY)**: PMFBY provides crop insurance coverage to farmers against yield losses, crop damage, and other risks. It ensures financial support to farmers in the event of adverse climatic conditions, pests, or diseases affecting their crops. The scheme offers affordable insurance premiums and timely claim settlements, reducing the financial impact of production risks on farmers.

iv. **Pradhan Mantri Krishi Sinchai Yojana (PMKSY):** PMKSY promotes sustainable water management and irrigation practices. It provides financial assistance for the development of irrigation infrastructure, micro-irrigation systems, and water conservation initiatives. By enhancing access to irrigation and reducing dependence on rainfall, PMKSY helps mitigate financial risks associated with water scarcity and drought conditions.

v. **Farmers' Producer Organizations (FPOs):** The government supports the formation and strengthening of FPOs, which are collectives of farmers that enable them to collectively access inputs, credit, technology, and markets. FPOs empower farmers by reducing their dependence on intermediaries, enabling bulk procurement, and improving bargaining power. This helps in better price realization and mitigates financial risks associated with market uncertainties.

vi. **Pradhan Mantri Kisan Samman Nidhi (PM-KISAN):** PM-KISAN is a direct income support scheme that provides financial assistance to small and marginal farmers. Under this scheme, eligible farmers receive direct income support of a fixed amount per year in their bank accounts. The scheme aims to provide financial stability and support farmers' income requirements.

vii. **Micro Irrigation Fund (MIF):** The MIF provides affordable loans to farmers for the adoption of micro-irrigation systems, such as drip irrigation and sprinkler irrigation. These systems improve water-use efficiency, crop productivity, and income potential for farmers. The fund offers concessional interest rates and financial assistance to promote the widespread adoption of micro-irrigation techniques.

viii. **Debt Relief Schemes:** The government periodically introduces debt relief schemes to alleviate the burden of agricultural debt among farmers facing financial distress. These schemes involve debt restructuring, rescheduling, or waiver of loans to provide relief to farmers and ensure their financial stability. The eligibility criteria and coverage of such schemes vary across states and regions.

These efforts and schemes by the Indian government aim to enhance farmers' financial resilience, reduce debt burden, and provide them with the necessary financial support for agricultural activities. Farmers can approach their respective state agricultural departments or financial institutions to access these schemes and avail themselves of the benefits.

Conclusion

In conclusion, risk management and mitigation are critical components of ensuring the long-term success and sustainability of agribusiness ventures. The inherent uncertainties and vulnerabilities in agriculture necessitate a proactive approach to identify, assess, and address potential risks. Through comprehensive risk assessment, agribusinesses can gain a deeper understanding of the various risks they face, including market volatility, climate change, pests and diseases, natural disasters, policy changes, and financial uncertainties. This knowledge allows them to prioritize risks based on their likelihood and impact, enabling informed decision-making. Effective risk mitigation strategies are then implemented to minimize the adverse effects of identified risks. Diversification helps spread risks by investing in multiple crops, markets, and regions. Insurance coverage and risk transfer mechanisms provide financial protection against crop failure, extreme weather events, and market fluctuations. Embracing technology and innovation enhances productivity and resource efficiency while mitigating production and operational risks. Collaboration and partnerships provide access to resources, expertise, and market opportunities, while also sharing risks and responsibilities. Contingency planning ensures prompt responses to unforeseen events, and monitoring systems and early warning mechanisms enable proactive decision-making. Financial management practices, including prudent budgeting, cash flow management, and risk hedging, help safeguard against financial risks arising from market fluctuations. These strategies collectively contribute to building resilience, adaptability, and sustainability in agribusiness operations. By effectively managing risks, agribusinesses can protect investments, optimize performance, and seize opportunities for growth. Risk management practices not only protect against potential losses but also create a foundation for innovation, efficiency, and competitiveness in the agricultural sector. To navigate the dynamic and volatile agricultural landscape, agribusinesses must continuously assess and reassess risks, adapt strategies, and remain vigilant. By integrating risk management into their business operations, agribusinesses can confidently navigate uncertainties, enhance their resilience, and position themselves for long-term success in the agri-food industry.

References

Alizadeh, Amir and N. Nomikos (2005). Agricultural reforms and the use of market mechanisms for risk management. Study commissioned by the Futures and Options Association. London. Cass Business School and FOA. March 2005.

Auer, Josef (2003). Weather derivatives heading for sunny times, Deutsche Bank Research. Current Issues. Frankfurt.

Balcombe, Kevin (2009). The nature and determinants of volatility in agricultural prices. Technical Report, report to the FAO.

Bielza Diaz-Caneja, Maria, C.G. Conte, F.J. Gallego Pinilla, J. Stroblmair, R. Catanero and C. Dittman (2009). Risk management and agricultural insurance schemes in Europe. Joint Research Centre. European Communities. Luxembourg.

Bielza Diaz-Canela, Maria, C.G. Conte, R. Catanero and F.J. Gallego Pinilla (2008). Agricultural Insurance Schemes II: index insurances. Joint Research Centre. European Communities.

BirdLife International. EEB, EFNCP, IFOAM, WWF Proposal for a new EU Common Agricultural Policy. March 2010. Cafiero, Carlo, F. Capitanio, A.

Cioffi and A. Coppola (2007). Risk and crisis management in the reformed European agricultural policy.

FAO.1992. *Risk analysis in dryland farming systems*, by J. Anderson & J. Dillon, Farm Systems Management Series No. 2., Rome.

Fleisher, B.1990. *Agricultural Risk Management*, Lynne Rienner Publishers, Boulder, Colorado, USA.

Hardaker, J.B., Huirne, R.B.M. & Anderson, J.R.1997. *Coping with Risk in Agriculture*, CAB International, Oxford, UK.

USDA. 1997. Introduction to risk management: understanding agricultural risks, Risk Management Agency, United States Department of Agriculture (USDA), Washington, DC.

USDA.1999. *Managing risk in farming: concepts, research and analysis*, Agricultural Economic Report No. 774, by J. Harwood, R. Heifner, K. Coble, J. Perry & A. Somwaru, Market and Trade Economics Division and Resource Economics Division, Economics Research Service, United States Department of Agriculture (USDA), Washington, DC

5 Value Chain Analysis and Optimization in Agribusiness

Neha Thakur[1], Niyati Thakur[2*], Himanshi Verma[1] and Anshuman Klate[1]

[1]Department of Business Management, Dr. YSPUHF, Nauni, Solan HP (173230)
[2]Department of Social Sciences, Dr. YSPUHF, Nauni, Solan HP (173230)
Corresponding author: *thakniyati18ur@gmail.com*

Abstract

The agribusiness sector is an emerging sector in Indian as well as the global economy. The knowledge of value chain and optimization is of utmost importance to make any agribusiness a success, efficient, competitive and sustainable. Value chain analysis offers a comprehensive framework for evaluating and improving the various stages involved in the production and distribution of agricultural products. This chapter focuses on the importance of value chain analysis and optimization in agribusiness. It examines the concept of the value chain, which comprises of all the activities and processes that add value to agricultural products. These activities include input supply, production, processing, marketing, and distribution. By mapping out the value chain, businesses gain insights into the interdependencies, costs, and value-added activities within the system. Furthermore, the chapter emphasizes on the significance of value chain optimization to enhance efficiency and competitiveness. Optimization involves identifying bottlenecks, inefficiencies, and areas for improvement at each and every stage of the value chain process. This process enables agribusinesses to streamline operations, reduce costs, increase productivity, and ultimately deliver better products to the consumer. In conclusion, value chain analysis and optimization are critical tools for agribusinesses. By examining and enhancing the various stages of the value chain, businesses can identify opportunities for efficiency gains, cost reduction, and increased value creation. Embracing value chain optimization in agribusiness contributes not only to the success of individual enterprises but also to the overall sustainability and growth of the agricultural sector as a whole.

Keywords: Optimization, Value Chain, Input supply, Agriculture, Efficiency, Sustainability and Agribusiness.

Introduction

Agriculture plays a crucial role in India, providing income and livelihood opportunities to a large portion of the population. It is considered the backbone of the country, and any development intervention in this sector directly contributes to the economy and the upliftment of rural communities. Over the years, Indian agriculture has undergone significant transformations, moving from subsistence farming to commercial agriculture. The country has achieved self-sufficiency in food production and a good degree of stability. However, despite these achievements, there are still several challenges that farmers face. These include operating on a smaller scale, weak information and communication linkages with markets, exploitation by middlemen, lack of adequate financial support, improper post-harvest management, and disorganized value chains or marketing channels. To address these issues, efforts can be directed towards improving marketing systems and adopting a value chain approach. This approach helps understand the dynamics of various operations along the chain, adding value at each stage. By strengthening market linkages, providing farmers with timely information, promoting direct interactions between farmers and consumers, and investing in infrastructure and post-harvest management, agriculture can become a more profitable business for farmers in India. Policy interventions, such as creating farmer producer organizations, promoting contract farming, and improving rural infrastructure, can also contribute to the development of a more efficient and profitable agricultural sector. The agricultural value chains in India often suffer from fragmentation due to the presence of multiple middlemen, a lack of information about other links in the chain, and inefficiencies throughout the entire process. This leads to lower incomes, particularly for those at the lower end of the chain. To address these issues, efficient management and organization of operations along the value chain framework have been recognized as strategies to improve efficiency. Value chain management involves organizing a series of connected activities that create value by transforming raw materials into goods or services that are purchased by consumers. It serves as a heuristic, analytical, and strategic tool for identifying and examining the different actors within a value chain. This approach helps to understand the dynamics of roles and operations, reward and distribution mechanisms, power structures, linkages, and knowledge transfer.

In today's world, food loss and waste is a significant global issue with far-reaching consequences. It is estimated that roughly one-third of the food produced for human consumption is lost or wasted every year. This occurs at various stages of the food supply chain, including production, processing, distribution, and consumption. The impacts of food loss and waste are indeed multifaceted. Here are some key points:

Hunger and Food Insecurity: The world produces enough food to feed its entire population, yet millions of people still suffer from hunger and malnutrition. By reducing food loss and waste, we can redirect those resources to meet the nutritional needs of those who are food insecure.

Environmental Impact: Food production requires significant amounts of land, water, energy, and other natural resources. When food is lost or wasted, all these resources go to waste as well, resulting in unnecessary greenhouse gas emissions, deforestation, water depletion, and soil degradation. By reducing food loss and waste, we can mitigate these environmental impacts and promote sustainable resource management.

Climate Change: The food system is a significant contributor to greenhouse gas emissions. When food is lost or wasted, it releases methane, a potent greenhouse gas, as it decomposes in landfills. Moreover, the production of food that goes to waste also generates emissions throughout the supply chain. By minimizing food loss and waste, we can help reduce the carbon footprint of the food system and mitigate climate change.

Addressing food loss and waste requires collective efforts from various stakeholders, including governments, businesses, and consumers. Here are a few strategies that can be implemented:

1. Improved Infrastructure and Technology: Investing in better storage, transportation, and processing facilities can help minimize food loss during production and distribution. Advanced technologies, such as precision agriculture and smart packaging, can also play a role in reducing waste.

2. Redistribution and Food Recovery: Establishing systems to collect surplus food from farms, markets, restaurants, and households can redirect edible food to those in needs. Food banks, donation programs, and partnerships with non-profit organizations can facilitate this process.

3. Consumer Awareness and Behavior Change: Educating consumers about the impacts of food waste and providing practical tips to reduce waste at home can make a significant difference. Planning meals, proper storage,

and understanding date labeling can help individuals reduce food waste

4. Policy and Regulation: Governments can enact policies that encourage the reduction of food waste, such as setting targets, providing incentives, and implementing waste management regulations. They can also support research and innovation to find sustainable solutions.

By addressing food loss and waste, we can not only alleviate hunger and promote food security but also contribute to a more sustainable and resilient food system. It requires a comprehensive approach involving all stakeholders to achieve meaningful progress.

Value Chain Analysis:

According to Kaplinsky and Morris (2002), a value chain is "the entire range of activities that are required to bring a product or service from conception, through the intermediary phases of production, delivery to final consumers, and disposal after use." Value chain concept views production as simply one of many value-adding components in the chain, as opposed to the previous exclusive focus on production. Value chain concept emphasizes value adding activities at each stage.

Value chain analysis is a comprehensive and strategic tool that helps identify and analyze the various actors, roles, and operations within a value chain. It examines aspects such as the distribution of rewards, power dynamics, linkages, and knowledge transfer. By conducting a value chain analysis, inefficiencies within the chain can be identified, and measures can be taken to improve overall performance. In the context of agriculture and agribusiness, value chain management involves organizing a series of interconnected activities along the chain to enhance value creation, reduce inventories, and improve customer satisfaction. It emphasizes the need for effective communication and coordination among the different members of the value chain. The goal is to ensure that each member receives a fair profit and that the chain functions efficiently.

Efficient value chain management in agriculture and agribusiness require collaboration, information sharing, and cooperation among the various stakeholders involved. By improving communication and coordination, issues such as information gaps, delays, and power imbalances can be addressed. This, in turn, leads to a more equitable distribution of benefits and improved overall performance of the value chain.

By implementing efficient value chain management practices, agricultural systems can become more competitive, responsive, and profitable. It allows for better resource allocation, reduced waste, and improved decision-making. Ultimately, the aim is to create a value chain where all participants receive fair returns for their contributions, leading to enhanced sustainability and profitability in the agricultural sector.

Optimization

Resource saving refers to the practices and actions undertaken by economic agents to optimize the utilization of productive resources. It involves continuous improvements in resource exploitation and establishing economic mechanisms aimed at conserving production resources. These mechanisms encompass resource management, financing, incentives, and more. The strategy for resource saving is tailored to the specific needs and goals of various entities, including individuals, companies, industries, regions, and even entire states. Each entity formulates and implements its own resource saving strategy based on its unique circumstances and objectives.

Businesses play a vital role in executing resource saving strategies. They actively engage in activities that promote efficient resource utilization, such as adopting sustainable production practices, implementing energy-saving technologies, reducing waste, and optimizing resource allocation. By embracing resource-saving practices, businesses can enhance their operational efficiency, reduce costs, minimize environmental impact, and improve their overall competitiveness. Furthermore, businesses can also contribute to resource saving through collaboration and innovation. They can participate in industry-wide initiatives, research and development efforts, and knowledge sharing platforms to collectively identify and implement resource-saving solutions. By working together, businesses can drive systemic changes that lead to more sustainable and resource-efficient practices across entire sectors or value chains. To increase resource-saving effects within an agricultural enterprise, it is advisable to concentrate efforts and specialize in specific areas of agricultural production, leveraging advanced technologies. By focusing on 2-4 important directions, the enterprise can streamline its operations and allocate resources efficiently (Malik et al., 2001). This specialization allows the enterprise to develop deep expertise in those chosen areas, making informed decisions and implementing best practices. Additionally, the targeted allocation of resources ensures optimal utilization, reducing waste and enhancing cost-effectiveness. The enterprise can also invest in advanced technologies tailored to their specific agricultural directions, such as precision farming tools and data analytics, improving productivity while minimizing resource consumption. By concentrating on a limited number of directions, logistics and supply chain management can be streamlined, resulting in cost reduction, improved product quality, and customer satisfaction. Furthermore, specialization facilitates increased investment in research and development, fostering innovation and the development of sustainable practices. Nonetheless, careful market analysis is essential to select directions that align with market demand, resource availability, and the enterprise's capabilities, ensuring long-term viability and success.

Value Chain optimization: Value chain optimization and production optimization are interconnected processes that aim to reduce waste, increase profitability, and enhance overall efficiency within a company. The principles that apply to one often apply to the other as well. To optimize the value chain, it is crucial for companies to have accurate and comprehensive data about the various stages of the value chain. This data should encompass both historical and current information, including costs, availability of resources, throughput, and inputs. By having a thorough understanding of the value chain processes and factors that impact production, companies can identify and address inefficiencies.

Companies that have adopted industrial digitization practices and possess process simulations are already equipped with valuable sensing data related to energy usage and asset productivity. These data can be leveraged to gain insights into operational performance and identify areas for improvement. Additionally, purchase records that track supplier reliability and pricing can provide valuable information for optimizing the value chain. By analyzing the data collected from these various sources, companies can identify inefficiencies, bottlenecks, and areas of improvement within the value chain. They can then take appropriate measures to eliminate waste, streamline processes, and enhance overall performance. This optimization can lead to cost savings, improved productivity, and increased profitability. It is worth noting that value chain optimization requires a holistic approach, considering all aspects of the value chain, from sourcing and procurement to production, distribution, and customer service. Collaboration and communication among different stakeholders, both internal and external, are crucial for successful value chain optimization.

How can it enhance firm's reliability: Delivering on contracts is a crucial aspect of building and maintaining strong business relationships for any company. As part of value chain optimization, it is essential to assess the reliability and performance of suppliers in meeting their contractual obligations. This evaluation allows companies to identify consistent and trustworthy suppliers and understand if any shortages or disruptions in the value chain are due to factors beyond a vendor's control. By monitoring and analyzing supplier performance, companies can gain insights into the reliability of their suppliers. This includes assessing whether suppliers consistently fulfill their contractual commitments, meet quality standards, and deliver products or services on time. Understanding the track record of suppliers helps identify reliable partners who contribute to a smoothly functioning value chain. Overall, by utilizing accurate and comprehensive data, companies can optimize their value chains, identify and eliminate inefficiencies, and ultimately enhance profitability and competitiveness in the market. Furthermore, it is important to determine if any given contract truly provides the required value to maintain that agreement. Evaluating the performance of suppliers in relation to the contract terms helps identify if they are delivering the expected benefits, such as cost savings, quality improvements, or timely delivery. This analysis helps reliability managers make informed decisions about the continuation or modification of contracts, ensuring that value is consistently generated.

Optimizing the value chain also involves assessing the number of suppliers required to sustain production. By considering the performance and reliability of existing suppliers, companies can determine whether they have a sufficient number of reliable partners to meet production demands. If suppliers are consistently falling short of expectations, it may be necessary to seek additional suppliers or diversify the supply base to mitigate risks and maintain continuous production. Therefore, evaluating supplier performance, contract value, and the number of suppliers is crucial for effective value chain optimization. By identifying reliable suppliers, assessing contract benefits, and ensuring an adequate supplier base, companies can enhance their operational efficiency, reduce risks, and maintain a smooth flow of production within the value chain.

Tools for Optimization: Analyzing this data using advanced analytics techniques, such as machine learning algorithms, Artificial Intelligence etc. allow companies to identify patterns, detect inefficiencies, and uncover opportunities for improvement. It enables them to make data-driven decisions and take proactive measures to optimize the value chain. By extracting actionable insights from their own data, companies can effectively address bottlenecks, reduce waste, streamline processes, and enhance overall performance. The key to successful value chain optimization lies in the ability of management to act upon these insights. Empowering management with the necessary authority and resources enables them to implement changes and improvements based on the data-driven insights. This might involve reconfiguring production processes, adjusting inventory levels, optimizing logistics routes, or adopting new technologies and practices. Implementing value chain optimization through intelligent data usage offers several benefits. It allows companies to enhance operational efficiency, reduce costs, improve product quality, optimize inventory management, and respond more effectively to customer demands. By leveraging their own data and acting upon the insights derived from it, companies can stay ahead of the competition and achieve a competitive advantage in the market.

Moreover, as the adoption of plant digitalization technology and practices continues to increase, companies can leverage value chain optimization to maximize profits, reduce emissions, and enhance production. With every step of production digitized, the data becomes readily available for analysis and optimization. Given the complexity of supply and value chains, employing an industrial AI solution proves to be the most effective approach. Artificial intelligence (AI) has witnessed significant advancements, offering powerful and user-friendly predictive tools that can greatly benefit companies. Industrial AI eliminates much of the guesswork involved in assessing the costs and benefits of variables within a company's supply chain. It enables companies to analyze large volumes of data, identify patterns, and uncover valuable insights. By leveraging AI, companies can make informed decisions based on data-driven predictions, optimizing their value chain for improved efficiency and profitability. One of the key advantages of industrial AI is its ability to isolate the business aspects of relationships with suppliers. By analyzing data related to deliverables and performance, AI systems can identify patterns and anomalies that might go unnoticed by humans. These systems can detect subtle variations or changes in the value chain, which could prove to be highly profitable insights for the company. By leveraging AI's analytical capabilities, companies can gain a competitive edge and capitalize on opportunities that may not be apparent through traditional analysis methods.

Furthermore, industrial AI can assist in real-time monitoring and continuous optimization of the value chain. It can identify inefficiencies, bottlenecks, and areas for improvement, allowing companies to take prompt action and make necessary adjustments. This proactive approach helps companies to stay agile, respond to market changes, and maintain a competitive position in their industry.

So, it can be concluded that the smart enterprise understands the value of data and its role in optimizing the value chain. By collecting, analyzing, and utilizing measurable data, companies can make informed decisions, drive operational improvements, and achieve greater efficiency and profitability throughout their value chain. Also, the industrial AI, powered by advanced data analytics and predictive tools, enhances value chain optimization efforts. It enables companies to leverage digitized data, uncover valuable insights, and make informed decisions. By leveraging AI, companies can maximize profits, reduce emissions, increase production, and uncover hidden opportunities within their value chain.

Optimization features: There are three main features of value chain optimization

1. **Agility:** A digital twin of the value chain is a virtual replica that provides

real-time operational insight. It enables companies to stay competitive in volatile markets by offering visibility into the current state of the value chain. With this agility, companies can make informed decisions, optimize operations, and adapt to market dynamics. The digital twin facilitates scenario planning, predictive analytics, collaboration, and communication across teams. It empowers companies to respond promptly to changes, reduce risks, and maintain a competitive edge.

2. **Sustainability:** Gaining insight into the carbon intensity of feedstock and emissions information from different operating conditions is crucial for agribusiness companies to achieve their sustainability goals. This data helps agribusiness companies make informed decisions, optimize processes, manage their supply chains, and drive innovation towards reducing their carbon footprint. By analyzing carbon intensity and emissions information, agribusiness companies can choose sustainable feedstock options, optimize their operations, track progress, and invest in innovative solutions for a more sustainable future.

3. **Innovative technologies:** Incorporating innovative technologies and leveraging AI capabilities can accelerate computations, simplify deployment, and uncover hidden insights. By utilizing high-performance computing, cloud platforms, and advanced AI algorithms, companies can solve complex problems faster, deploy solutions more efficiently, and benefit from AI-driven recommendations. These technologies enhance computational capabilities and drive efficiency, innovation, and competitive advantage.

Value Chain Analysis in Agribusiness

The agricultural value chain acts as a road map for comprehending how the sector operates and making the most of resources. Because of this, it's crucial for organizations to understand what goes into each stage of a value chain. Small farmers, who produce items in very small amounts and make upto 95% of those working in the agriculture industry, are unable to compete with the importers of goods. Value chain analysis may be performed at two different geographical and product levels, defining input-output flows within a specified economy and measuring input-output flows based on a defined functional unit of a commodity without being site-specific. Typically, there are two main categories into which agricultural cooperatives fall. Cooperatives for the production of commodities in which the members share ownership of and use of the production factors (land, buildings and structures, equipment, mechanisation, etc.). The most common type of cooperatives worldwide is service cooperatives. The members of such associations do not combine the properties and production elements. In contrast to before, when they each independently created retail demand, the cooperative enables its members to jointly create wholesale demand. It also enables members to consolidate their products, process them, and sell them together. It also enables members to engage in joint marketing activities.

Input Supply: The beginning of a typical agricultural value chain starts with the input supply. Similar to the pre-production phase, it contains strategies for genetic resource conservation and utilization to raise agricultural output and product quality. It is also where new developments in agricultural equipment, seeds, irrigation, and fertilizer are discussed. For a better harvest, more revenue, and to ensure that farmers can prosper in the face of environmental challenges like climate change, it is crucial to take these factors into account early in the production process.

Production: Some of the conditions that can be controlled and some that cannot are present during the production stage of planting, growing, and harvesting. Most businesses also prepare backup plans in case of pest and disease outbreaks, erratic weather, shifting market conditions, and food loss management by foreseeing these issues directly through an agricultural value chain. With so many outside influences on the agricultural sector, it may be unpredictable at times, therefore it's important to address such issues.

Trading: At this point, strategies for getting food from farmers to consumers are being considered. Some items may be processed, packed, and stored before being given to customers since not all agricultural products can be delivered directly to the market. In trading, product quality is also assessed and valued. This stage addresses post-harvest food loss problems raised by shoddy road systems that may create delays in the transportation of supplies. To guarantee the finest quality once it is on the market, some products need to be shipped immediately away.

Warehousing: Another crucial phase of the agricultural value chain is storage. It entails keeping products safe and secure from the point of manufacture till consumption. Perishable commodities' quality is protected from degradation by storage. Continuous manufacturing and storage are required for goods with seasonal demand.

Processing: After harvest, products are refined at this step to get them ready for on-site processing and packing. In addition to packing, you may also sort and grade agricultural products, plant products, or livestock into finished items for consumption.

Wholesale or Export: Agricultural wholesaling and exports are frequent, but they need careful planning since they entail moving enormous amounts of products to markets and even outside of the nation of origin. Farmers and businesses must deal with customs when their crops, animals, or plant-based goods arrive in the destination country in addition to assessing their quality.

Retail: This entails making comparatively lower sales of agricultural products. Instead of being sold through resellers, retail items often go directly to customers.

Case Study: Giuli Keshelashvili (2018) analyzed the Agricultural Cooperatives' Management in Georgia. Georgia had 1586 recognised agricultural cooperatives as of 1 January 2017, and they are spread throughout the regions like 156 in Kakheti, 224 in Kvemo Kartli, 86 in Mtskheta-Mtianeti, 119 in Shida Kartli, 127 in Samegrelo-Zemo Svaneti, 53 in Guria, 155 in Adjara, 358 in Samtskhe-Javakheti, 124 in Imereti, 132 in Racha-Lechk. The distribution of cooperatives is most evenly spread in the Racha-Lechkhumi and Kvemo Svaneti areas, while it is most unequal in the Samtskhe-Javakheti region. The majority of the potatoes were grown by the cooperatives. The survey, which fell within the purview of our study, revealed that for the cooperatives operating in Georgia in 2017, high incomes, production growth, employment, improvement of product quality, and improvement of raw material, supply, infrastructure, and market availability were the most significant benefits. Qualitative research techniques to analyze the administration of agricultural cooperatives particularly focus group sessions and interviews that were done to identify difficulties. The following topics were covered during the study process: management organizational structure, decision-making and communication processes, organizational strategies, employee professionalism and qualifications, and staff development. The study concluded that by better involving themselves in the sales process, cooperative formation would enable farmers to capitalize on market possibilities and lower costs at numerous points along the value chain. The establishment of cooperatives at the second and third levels in addition to the first level would further streamline the value chain, lower expenses, and enhance advantages for primary producers, therefore promoting the expansion of their operations.

Optimization: Optimization refers to the process of finding the best possible solution or outcome from a set of available options or constraints. It involves maximizing desired factors or minimizing undesired ones to achieve the most favorable result. In various fields such as mathematics, engineering, economics, and computer science, optimization techniques are used to optimize objectives, improve efficiency, allocate resources effectively, and solve complex problems. Optimization involves analyzing and evaluating different possibilities, considering various variables and constraints, and iteratively refining the solution until an optimal or near-optimal outcome is achieved. (Yakov Ben-Haim 2012). In the context of profit and cost, optimization refers to the process of maximizing profit or minimizing cost by identifying the best allocation and utilization of resources. It involves analysing various factors such as production levels, pricing strategies, resource utilization, and operational efficiencies to achieve the optimal balance between revenue generation and cost management. When optimizing for profit, businesses aim to maximize their earnings by identifying the most effective strategies for increasing revenue and reducing expenses. This may involve optimizing pricing strategies to maximize profit margins, identifying cost-saving opportunities in the supply chain or production process, and identifying the most profitable product mix or customer segments. (Spall J.C 2003) On the other hand, optimizing for cost involves minimizing expenses while maintaining desired levels of productivity and quality. This may involve identifying cost-saving measures such as streamlining processes, improving operational efficiency, negotiating better supplier contracts, or identifying alternative cost-effective resources. In both cases, optimization requires careful analysis, modelling, and decision-making based on quantitative and qualitative data. Various optimization techniques, such as linear programming, nonlinear programming, or simulation, can be employed to find the optimal solution that maximizes profit or minimizes cost, considering the constraints and objectives of the specific business or scenario.

Optimization in Agribusiness

The emergence of increased specialization and the adoption of capital-intensive production systems have driven the need for more formal planning methods in agriculture. While traditional judgment based on experience has long been relied upon, the development of mathematical models has provided a quicker and less expensive alternative. By constructing and analyzing these models, agricultural planners can derive and test solutions, leading to more efficient decision-making. Implementing and monitoring these solutions allows for performance evaluation and control, ultimately enhancing productivity in agricultural systems. The utilization of mathematical modeling offers a valuable tool that complements and enhances traditional planning approaches in agriculture. Decision making has become an integral part of the daily responsibilities for individuals in technical, managerial, or administrative roles. Whether it pertains to ensuring the right composition of a mixture, selecting suppliers, optimizing production quantities, allocating resources, making investment choices, planning routes, optimizing land utilization, or staffing checkout lanes, decision-making is ubiquitous across various professional domains. The ability to make effective decisions, considering factors such as specifications, available options, demands, and constraints, is crucial for success in today's dynamic work environments. Embracing thoughtful and informed decision-making processes empowers professionals to navigate challenges, optimize outcomes, and drive efficiency in their respective fields. a situation that necessitates decision making is commonly referred to as a decision-making problem. These problems arise within a specific system, known as the relevant system, where decisions must be made. The individuals responsible for making these decisions are referred to as decision makers. Decision problems can range from simple scenarios with a few controllable variables and straightforward conditions to large-scale, complex situations with numerous variables and multiple conditions to be met. Regardless of the complexity, decision making always entails selecting the most suitable option among a range of alternatives. It is a process that requires careful consideration, analysis, and evaluation of available choices to determine the optimal course of action (Eppen et al. 1993).

In recent times, farm planning has gained significant complexity and importance, particularly in developed regions, where agriculture has transitioned toward larger and more specialized farm units. These transformations have acted as catalysts for the advancement of formal planning techniques that rely on mathematical models. While these techniques bear resemblance to operations research, the direct engagement of the profession in agricultural planning has remained somewhat limited, with a considerable portion of published work being associated with agricultural economics. The increasing scale and specialization of farms have created a demand for more sophisticated approaches to address the intricacies of modern agricultural systems. Mathematical models have emerged as valuable tools in farm planning, enabling farmers to optimize resource allocation, production strategies, and risk management. These models provide a systematic framework for decision-making, allowing for a comprehensive analysis of various factors such as crop selection, input utilization, market conditions, and financial considerations. While the involvement of operations research in agricultural planning may not be as extensive as in other fields, the contributions from agricultural economics have played a crucial role in advancing planning techniques. The interdisciplinary nature of farm planning calls for collaboration among professionals from various disciplines, including agricultural economics, operations research, agronomy, and environmental sciences, to effectively address the complex challenges faced by modern farming operations. As agriculture continues to evolve, the integration of formal planning methods based on mathematical models, coupled with insights from agricultural economics and related disciplines, will likely play an increasingly important role in optimizing farm operations, improving sustainability, and ensuring the long-term viability of agricultural systems. (Yang, 1995).

Indeed, optimization is closely intertwined with decision making. It encompasses a set of techniques that facilitate the process of arriving at optimal or best decisions. The concept of optimization has become deeply ingrained as a fundamental principle in the analysis of numerous intricate decision or allocation problems. Its application brings forth a certain degree of philosophical elegance that is difficult to challenge, while simultaneously offering operational simplicity that proves invaluable in practice. Optimization techniques aim to find the most favorable outcome, considering a range of variables, constraints, and objectives. By formulating problems as mathematical models and employing optimization algorithms, decision makers can systematically explore the solution space and identify the optimal or near-optimal solution. Whether maximizing profit, minimizing cost, optimizing resource allocation, or achieving desired performance levels, optimization provides a powerful framework for informed decision making. The beauty of optimization lies in its ability to handle complex problems with clarity and efficiency. It simplifies decision-making processes by providing quantitative methods to evaluate alternatives, assess trade-offs, and identify the best course of action. Optimization allows decision makers to navigate the intricacies of real-world scenarios, taking into account multiple variables and constraints, and arriving at solutions that maximize desired outcomes. While acknowledging the philosophical elegance of optimization, it is important to recognize that its application is not without challenges. Real-world complexities, uncertainty, and limited information can introduce additional layers of complexity. However, optimization remains a valuable tool for guiding decision making, offering a systematic approach to tackle complex problems and find optimal solutions that align with objectives and constraints. In summary, optimization serves as a powerful ally in decision making. It provides a framework for identifying optimal or best decisions, contributing both philosophical elegance and operational simplicity. By leveraging optimization techniques, decision makers can navigate complex problem domains, explore the solution space, and ultimately make informed choices that maximize desired outcomes.

Optimization and optimization model

Optimization is a mathematical discipline that focuses on finding the maximum or minimum values of functions, often subject to constraints. It is also referred to as mathematical programming, where the objective is to select the best element from a set of available alternatives based on certain criteria. The essence of optimization lies in achieving the best possible outcome given specific circumstances. In decision-making processes, the goal is typically either to minimize effort or to maximize benefit. This objective can be quantified as a function that depends on various design variables. Optimization, therefore, involves identifying the conditions or values of these variables that yield the maximum or minimum value of the function. Throughout history, humans have demonstrated a consistent interest in optimizing the performance of the systems they use. In contemporary times, both professional and personal decisions revolve around the pursuit of optimizing desirable characteristics. When there are multiple objective functions to optimize, alongside the need to satisfy constraints on decision variables, the resulting model is known as an optimization model (Jayachandran,1985). Optimization models offer a framework for systematically analyzing and solving problems to attain optimal outcomes. They allow for the identification of optimal strategies, configurations, or decisions that align with specific objectives and constraints. By leveraging mathematical techniques and algorithms, optimization helps individuals and organizations make informed choices and improve efficiency in various domains. In summary, optimization is a mathematical discipline that strives to achieve the best possible outcome by finding the maximum or minimum values of functions. It plays a central role in decision making, enabling the optimization of desirable characteristics and the fulfillment of requirements. Optimization models provide a structured approach to problem-solving, guiding individuals and organizations toward optimal solutions in diverse contexts.

Case study: The study conducted by Sonmez and Altil in 2004 focused on developing a linear programming model for irrigation scheduling and cropping patterns in Harrran plain, Turkey. The aim was to explore the impact of both adequate and deficit water supplies on farm income. The findings of their research revealed that even under conditions of very low water supply, it was still feasible to maintain farm income at high levels. By utilizing a linear programming model, the researchers were able to optimize irrigation scheduling and determine the most suitable cropping patterns considering different water supply scenarios. The model allowed for the evaluation of various factors, including crop water requirements, available water resources, and market conditions, to identify the optimal allocation of water resources across different crops. The results demonstrated that effective irrigation scheduling and appropriate cropping patterns could mitigate the negative effects of water scarcity on farm income. Despite facing limited water supplies, the study showcased that careful planning and management can help farmers maintain profitability by optimizing resource utilization and aligning cropping patterns with market demands. The research conducted by Sonmez and Altil highlights the significance of adopting efficient irrigation strategies and crop selection techniques in regions with water scarcity. By employing mathematical modeling and optimization techniques, farmers and policymakers can make informed decisions that contribute to sustaining agricultural productivity and income levels, even in challenging water supply conditions. Overall, this study provides valuable insights into the potential for maintaining high farm income in situations of low water supply, emphasizing the importance of effective water management and crop planning in ensuring the economic viability of agricultural systems.

Conclusion

Value chain analysis and optimization are both valuable tools in strategic management and decision-making within businesses, including agribusiness. Value chain analysis helps organizations understand the sequence of activities involved in delivering a product or service to customers. By examining the primary and support activities within the value chain, businesses can identify areas where they can create value, differentiate themselves from competitors, and improve overall performance. This analysis provides insights into cost drivers, competitive advantages, and opportunities for innovation and improvement.

Optimization, on the other hand, focuses on maximizing efficiency, productivity, and profitability within specific areas of the business. In agribusiness, optimization can be applied to various aspects such as crop production, livestock management, supply chain logistics, and resource allocation. By utilizing optimization techniques, businesses can make strategic decisions to maximize resource utilization, minimize costs, reduce waste, and increase overall performance and profitability. When applied together, value chain analysis and optimization complement each other. Value chain analysis helps identify the activities and processes that are critical to delivering value, while optimization techniques provide the means to improve and optimize those activities. By leveraging the insights gained from value chain analysis, businesses can prioritize areas for optimization and implement strategies to enhance performance, sustainability, and competitiveness. Ultimately, the combination of value chain analysis and optimization enables businesses, including those in agribusiness, to better understand their operations, identify areas for improvement, and make informed decisions that lead to enhanced efficiency, profitability, and long-term success.

References

Benhaim Y. 2012. Doing our best: optimization and the management of risk. *Risk Analysis: An International Journal*, **32**(8), 1326-1332.

Eppen GD, Gould FJ and Schmidt CP. 1993. *Introductory Management Science (* 4th Ed.). Upper Saddle River, N.J.:Prentice-Hall.

Fabe A, Grote U and Winter E. 2009. Value chain analysis methodologies in the context of environment and trade research (No. 429). Discussion papers//School of Economics and Management of the Hanover Leibniz University, Germany.

Giuli keshelashvili. 2018. Value Chain Management in Agribusiness. *International Journal of Business and Management,* **6**(2), pp. 59-77., 10.20472/BM.2018.6.2.004

Jayachandran N. 1985. Optimization of enterprise combinations with special reference to garden land agriculture. M.Sc. (Ag) thesis, Kerala Agricultural University, Thrissur. 82p.

Malik Mykola Io, Zinovchuk Vitalii V and Sabluk Petro T. 2001. Basics of Agricultural Entrepreneurship, Editor Malik M.Io. – Kyiv. 582 p. (in Ukrainian).

Sonmez FK and Altil M. 2004. Irrigation schedule and optimum cropping pattern with adequate and deficit water supply for mid-sized farms of Harren Plain. *Pakistan J. Biol. Sci.* 7(8): 1414-1418.

Spall JC. 2003. *Introduction to Stochastic Search and Optimization, Estimation, Simulation and Control, Wiley Publications*, 595 p.

Yang WY. 1995. *Methods of Farm Management in Investigation for Improving Farm Productivity*. FAO, Rome, 228p.

6 Farm Management Principles

Dr. Sumit B. Wasnik[1] and Dr. Sneha Pandey[2]

[1]Assistant Professor (Guest Faculty), RSV CARS, Bemetara (CG)
[2]Assistant Professor (Guest Faculty), Rajmata Vijyaraje Scindia Krishi Vishwa Vidyalaya, Gwalior (MP)

Abstract

Nobody can deny the fact that it is the endeavor of the farming community only that helps realize the higher farm production. Farm management, includes making and implementing of the decisions involved in organizing and operating a farm for maximum production and profit. Farm management draws on agricultural economics and fall under microeconomics. In this chapter we studied about eight principles of farm management viz., comparative advantage, opportunity cost, theory of cost, product substitution, factor substitution, diminishing marginal return, equi-marginal returns and time comparison. These principles provide the farmer to find the practical solution of farm related problems such as production, marketing as well as management of on farm resources. Along with efficient resource management these principles guide to achieve profit maximization level at farm. Thus, farm management tools help farmer in solving farm related problems for successful farm business.

Keywords: Costs, Diminishing Return, Farm management, Input, Principles, Rational, Output.

Introduction

The prosperity of any country depends upon the prosperity of farmers, which in turn depends upon the rational allocation of resources among various uses and adopting improved technology. To enable the farmers to gear up the situation, they need to have knowledge on concept, principles and various issues of the farm to manage. The role of farm management, is therefore, to supply the information for the farmers for sound planning. All farm management tools are helpful to the farmers in solving their managerial problems for successful operation on the farm business.

In the context of increased commercialization there is a greater need to improve the managerial abilities of the farmer. So far, the managers in general have responded admirably to the technological changes that accrued in Indian agriculture. Human race depends more on farm products for their existence than anything else since food, clothing – the prime necessaries are products of farming industry. Even for industrial prosperity, farming industry forms the basic infrastructure. Thus, the study farm management has got prime importance in any economy particularly on agrarian economy.

Meaning of Farm Management

Farm Management comprises of two words i.e., Farm and Management. Farm means a piece of land where crops and livestock enterprises are taken up under common management and has specific boundaries. It is also called as socio-economic unit which not only provides income to a farmer but also a source of happiness to him and his family. While, management is the art of getting work done out of others working in a group. Thus, management is the process of designing and maintaining an environment in which individuals working together in groups accomplish selected aims. Hence, it is the key ingredient. The manager makes or breaks a business. He plays important role in making various farm related decisions.

Definitions

1. Farm management is a branch of agricultural economics, which deals with business principles and practices of farming with an object of obtaining the maximum possible return from the farm as a unit under a sound farming programme (Raju and Rao).

2. Farm management is a branch of agricultural economics which deals with wealth earning and wealth spending activities of a farmer, in relation to the organization and operation of the individual farm unit for securing the maximum possible net income (Bradford and Johnson).

Farm management is considered to fall in the field of *microeconomics*. Its objective is to maximize returns from the farm as a whole as it is interested in profitability along with practicability. What crops and livestock enterprises and their combinations to grow, what number of resources to be applied, how the various farm activities to be performed, etc., all these falls within the scope of farm management.

Nature of Farm Management

Farm management deals with the business principles of farming from the point of view of an *individual farm*. Its field of study is limited to the individual farm as a unit and interested in maximum possible returns to the individual farmer. It applies the local knowledge as well as scientific finding to the individual farm business. Farm management in short be called as a *science of choice or decision making*.

In other words, farm management seeks to help the farmer in deciding problems. The farm manager has to take decisions on several aspects for profitable operation of the farm. Decisions have to be taken regarding production (what to produce?, when to produce?, how much to produce?, how to produce?, etc.), marketing (buying and selling) and Organization and managerial i.e., administration. These three segments are interlinked and decision making is also interlinked.

Thus, the objectives to be fulfilled for successful farm management are: (i) examination of production pattern and resource use on the farm; (ii) identification of the responsible factors for present production pattern; (iii) determination of the optimality conditions in resource use and production; (iv) analysis of sub-optimality condition in the resource use and (v) provide suggestions in optimal use of resources at present.

Principles of Farm Management

A knowledge of economics provides a decision maker with a set of principles for decision making which are useful when preparing farm plans to organized a farm business. The economics principles guide the manager to setting the goals and preparing plans based on optimum resource allocation, resource substitution and combination and enterprise combination. Basically, farm management is the application of agricultural sciences and economic principles to the organization and operation of a farm business. The economic principles guide the manager in setting the goals and preparing plans based on optimum resource allocation, resource substitution and enterprise combination.

The outpouring of new technological information is making the farm problems increasingly challenging and providing attractive opportunities for maximizing profits. Hence, the application of economic principles to farming is essential for the successful management of the farm business. Some of the economic principles that help in rational farm management decisions are;

1. Principle of Diminishing Marginal Returns

Marshall provided the foundation for the economic principles of farm management through his law of diminishing returns. The law of diminishing returns is a basic natural law affecting many phases of management of a farm business.

According to Marshall ***"An increase in the capital and labour employed in the cultivation of land cause, in general, a less than proportionate increase in the amount of produce raised unless it happens to coincide with an improvement in the art of agriculture".***

Stages of Law of Diminishing Returns

Three components are considered into law are total product, marginal product and average product.

Table 6.1 shows three stages of the law, the stages are:

1. Increasing return
2. Constant returns
3. Diminishing returns

Table 6.1: Total product, Marginal product and Average product

Units	Total Product	Marginal Product	Average Product	Remarks
0	0	-	-	Stage-I Increasing Returns
1	6	6	6	
2	14	8	7	
3	27	13	9	
4	39	12	9.75	
5	50	11	10	Stage-II Constant Returns
6	60	10	10	

7	60	0	8.57	Stage-III Decreasing Returns
8	58	-2	7.25	
9	54	-4	6	

Stage I: In this stage, AP and MP are increasing, but MP is always greater than AP. Hence more units of input increase production in greater proportion. In this stage the input used are so small comparative to the fixed input. For profit maximization, produce more by using more units of input.

$$\frac{\Delta Y}{\Delta X} > \frac{Px}{Py}$$

Inflection Point: It is the point where TP curve changes its curvature and at this point the slope of MP is maximum.

This stage is also called as *irrational stage*. Because the efficiency of all the variable inputs keeps on increasing throughout this stage as indicated by increasing AP (technical efficiency). So, it is not reasonable to stop using an input when its efficiency is increasing and if we stop, some of the resource will remain unused or underutilized. Therefore, it is advised that farmer should continue to produce up to the level where AP is maximum.

Stage II: Stage II is the relevant stage of production, because the marginal product, though decreasing is not negative. A point where the optimum value of MP equals to its price is the condition of profit maximization. In this stage both the variable input and fixed input are being used in efficient amount relative to each other.

$$\frac{\Delta Y}{\Delta X} = \frac{Px}{Py}$$

This stage is also called as *rational stage* because of optimum utilization of inputs used in the production as remarked by increasing TP (point of congestion). In between this stage the technical efficiency of variable and fixed inputs is maximum. In this stage farmers are able to produce maximum output per unit of input. The objective of full resource use and maximum output is achieved. Therefore, farmers are advised to produce in this stage only.

Stage III: In this stage, total production is decreasing and MP is less than zero. There is over use of input relative to fixed input. This stage is not profitable for production as a farmer get less and less product for increasing amounts of input.

$$\frac{\Delta Y}{\Delta X} < \frac{Px}{Py}$$

This stage is also called as *irrational stage* because in this zone the TP is decreasing and MP becomes negative thereby technical efficiency of variable and fixed resources declines which indicates that additional quantities of inputs reduce the total output. It means that the inputs were overutilized. So, it is not advisable to operate in this region even if the additional quantities of inputs are available free of cost. Therefore, farmer should not operate in this stage otherwise he will have found reduced production as well as bear unnecessarily additional cost of inputs.

Production Function for input-output relationship (Short-run production function):

$$Y = f(X_1 | X_2 X_3 X_4 \dots\dots\dots X_n)$$

Where, Y = Output;

X_1 = Variable input;

$X_2\, X_3\, X_4 \dots\dots\dots Xn$ = Fixed inputs;

| (Vertical Bar) = It separates variable inputs from fixed inputs.

Reasons for the Operation of the Law of Diminishing Returns in Agriculture

1. **Excessive Dependence on Weather:** A farmer, however good he is in managing the farm, may not get the expected yields as he has little control over weather. A bad weather is just enough to make his expectation go topsy-turvy.
2. **Less Scope for Division of Labour:** There is no possibility of division of labour in farming as the farmer himself performs the role of labourer, manager and capitalist, therefore the advantage of division of labour is not a possibility. Therefore, the law of diminishing returns sets in quickly in farming.
3. **Less Scope for Mechanization:** Though mechanization of the farms enhances the productivity, they stand against using the machines for various farm operations. Under this limitation, the farmer fails to derive the advantage of mechanization.
4. **Cultivation of Inferior Lands:** To meet the food requirements of the teeming millions of populations, even inferior lands are brought under plough, the productivity of which in general is low.
5. **Continuous Cultivation:** Continuous cultivation of land drains out the fertility status

of the soil, thereby leading to low productivity. It is a fact that no farmer can afford to keep the land fallow for some period, to allow the land build up its fertility status.

2. Principle of Equi-Marginal Returns

In the previous principle, under the condition of unlimited resources, law of diminishing returns operates in determining the most profitable level of resource use but farmers have the limited resources i.e., limited land, limited capital, limited irrigation facilities. Even the labour which is considered to be surplus becomes scarce during peak sowing, weeding and harvesting periods. Under such resource limitations, farmers must decide how a limited amount of input should be allocated or divided among many possible uses or alternatives.

The equi-marginal principle provides guidelines for the ***rational allocation of scare resources*** in such a way that profit is maximized from each unit of input.

The principle says that ***returns from the limited resources will be maximum if each unit of the resource should be used where it brings greatest marginal returns.*** OR

A limited input should be allocated among alternative uses in such a way that the marginal value products (MVP or additional returns) of the last unit are equal in all its uses.

The law states that profits are maximized by using a limited resource in such a way, that the marginal returns from that resource are equal in all cases.

Table 6.2: Example- A farmer has Rs.5000 and wants to grow sugarcane, wheat and cotton that are suitable for his farm situation. What combination of amount of money should be spent on each enterprise to obtain highest profit?

		MVP		
Units of Amount spend	**Amount of money spent (Rs.)**	**Sugarcane (Rs.)**	**Wheat (Rs.)**	**Cotton (Rs.)**
1st	1000	2000	1900	2100
2nd	1000	1900	1800	1900
3rd	1000	1500	1500	1500
4th	1000	1200	1100	1200
5th	1000	1000	900	1100
Total returns (Rs.)	5000	7600	7200	7800
Net profit (Rs.)		2600	2200	2800

It is observed from the table that, when the entire amount was invented in any one crop, net profit from sugarcane, wheat and cotton is obtained as Rs. 2000, 1900 and 2100 respectively.

Table 6.3: However, if the same amount is spent according to principle of equi-marginal returns, total net profit will be as shown in the table given below;

Units of Amount spend	Amount (Rs.)	Crop	Marginal Return
1st	1000	Cotton	2100
2nd	1000	Cotton	1900
3rd	1000	Sugarcane	2000
4th	1000	Sugarcane	1900
5th	1000	Wheat	1900
Total (Rs.)	5000		9800
Net profit (Rs.)			4800

It is observed from the table 6.3 that, the farmer is getting total net profit of Rs.4800 which is more than profit from any single crop. Thus, for maximum net profit farmer should invest Rs.2000 in cotton, Rs.2000 in sugarcane and Rs.1000 in wheat.

Marginal return from all the three enterprises is equal i.e., Rs.1900. Thus, it can be invested in such a way that marginal returns should be in all the alternatives.

3. Principle of Opportunity Cost

An opportunity cost is the earning from the next best alternative sacrificed. Opportunity cost is the income from an input, which is used in a production process and at that particular time it has no alternative use. This means that the input will be losing from the alternative use and this income foregone by this input from its alternative use.

"Opportunity cost is the income that could be received, if the input had been used in its most profitable use."

Whatever must be given up to obtain some item is called opportunity cost. The opportunity cost is referred to as the real cost or true cost or alternate cost of an input. This, principle guides the farmer/manager that, *if the returns from the current use of the input are less than its opportunity cost, then the decision is to be changed.*

Example: If a farmer has Rs.1000 at his disposal, he has three alternatives i.e., investing on sugarcane, wheat and cotton. As given in table 6.2 Rs. 1000 is spent on sugarcane which is giving an MVP of Rs.2000. The farmer is forgoing the other two options by spending on cotton. Between the two, sugarcane is giving higher MVP than wheat. The farmer is sacrificing Rs.2000 from sugarcane, which is next best alternative to cotton, which is the opportunity cost.

4. Principle of Factor Substitution (Least Cost Combination)

This economic principle explains one of the basic production relationships viz., ***factor-factor relationship***. It guides in the determination of ***least cost combination of resources***. It helps in making a management decision of ***how to produce***. It implies the choice of methods of production or technology or the type of combination of resources given the constant output. This principle guides the farmer/manager in choosing the most appropriate combination of inputs that produces a given quantity of output with least cost. Substitution describes replacement of the process of production, exchange and consumption from less efficient to more efficient.

The principle of factor substitution says that *it is economical to substitute one resource (add resource) to another resource (replaced resource), as long as the reduction in the cost, resulting from decrease use of replaced resource is more than the increase in the cost due to increased use of added resource.*

In other words, the principle of factor substitution says that go on adding a resource so long as the cost of resource being added is less than the saving in cost from the resource being replaced. The principle of least cost combination of two inputs with a given level of output says that the marginal rate of substitution between inputs and inverse price ratio should be equal.

Production Function (Short-run production function):

$$Y^* = f(X_1, X_2 \mid X_3, X_4, \ldots\ldots\ldots Xn)$$

Where, Y^*= Constant Output level;

X_1 and X_2 = Variable input;

$X_3, X_4, \ldots\ldots\ldots Xn$ = Fixed inputs;

| (Vertical Bar) = It separates variable inputs from fixed inputs.

To find out least cost combination of resources:

Decrease in cost>Increase in cost

1. Marginal rate of substitution ($MRTS_{X1X2}$):

$$\frac{\Delta X2}{\Delta X1}$$

$\Delta X2$ = Number of units of replaced resource

$\Delta X1$ = Number of units of added resource

2. Inverse Price ratio (PR):

$$\frac{PX1}{PX2}$$

PX2 = Cost per unit replaced resource

PX1= Cost per unit added resource

Point where the substitution ratios and price ratios are equal:

$$\frac{\Delta X2}{\Delta X1} = \frac{PX1}{PX2}$$

Whether, X1 substitutes for X2 or X2 substitutes for X1, the profit rules:

1. If MRTS>PR (substitution ratio is more than inverse price ratio), one can reduce the costs by using more of "added" resource.

$$MRPS_{X1X2} = \frac{\Delta X2}{\Delta X1} > \frac{PX1}{PX2}\text{; increase X1}$$

2. If MRTS<PR (substitution ratio is less than inverse price ratio), cost can be reduced by using more of "replaced" resource.

$$MRPS_{X1X2} = \frac{\Delta X2}{\Delta X1} < \frac{PX1}{PX2}\text{; increase X2}$$

3. If the MRTS=PR (substitution ratio is equal to inverse price ratio), it is the point of least cost.

$$MRPS_{X1X2} = \frac{\Delta X2}{\Delta X1} = \frac{PX1}{PX2}$$

Example: Suppose there are six combinations of inputs. Find out the least cost combination.

Table 6.4: Tabular method

Combination	X1	X2	$MRTS_{X1X2}$	Inverse Price Ratio
A	75	325	4.34	2.25
B	**75**	**169**	**2.25**	**2.25**
C	75	125	1.67	2.25
D	75	122	1.62	2.25
E	75	90	1.20	2.25
F	75	85	1.13	2.25

From above combination, 75 units of X1 and 169 units of X2 is the least cost combination of inputs.

Table 6.5: Algebraic method

Combination	X1	X2	ΔX_1	ΔX_2	$MRTS_{X1X2}$	PR_{X1X2} (Inverse) = $\frac{P_{X_1}}{P_{X_2}}$
A	1	220	-	-	-	40
B	2	151	1	69	69	40
C	**3**	**111**	**1**	**40**	**40**	**40**
D	4	75	1	35	35	40
E	5	50	1	25	25	40
F	6	35	1	15	15	40
G	7	15	1	20	20	40

At Combination "C", = $MRTS_{X1X2}$ = PR is observed. So, this combination is the least cost combination.

Graphic method

Iso-quant curve: The curve representing all possible combinations of two inputs (X1 and X2) that produce the equal quantity of output is called an iso-quant.

Iso-cost line: the line represents all possible combinations of two inputs that can be purchased with the given amount of fund or outlay.

The intersection of Iso-quant curve with the iso-cost line is very important in the determination of the least cost combination of two inputs.

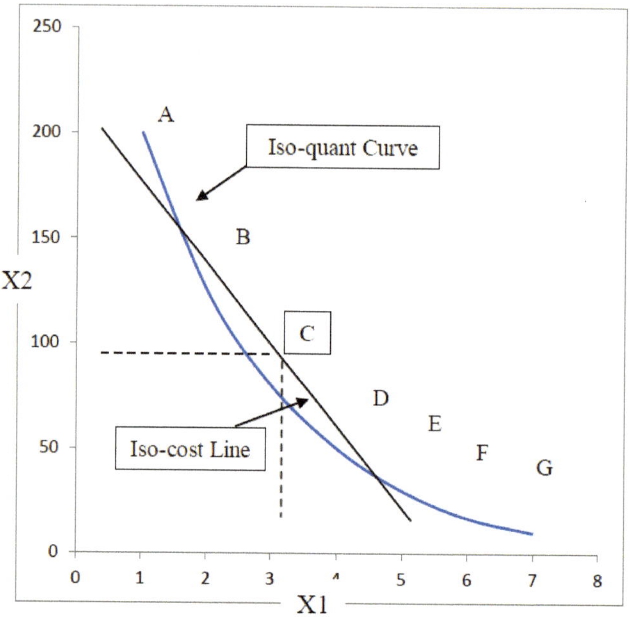

Combination "C" is found to be least cost because at this level Iso-quant curve and Iso-cost line intersects to each other.

5. Principle of Product Substitution

This principle involves ***product-product relationship*** which is associated with farmer's decision about what to produce? while, the objective of this principle is profit maximization.

This principle guides the producer in the determination of optimum combination of enterprises that maximize profits. If the inputs are constant, it is economical to substitute one product for the other, if the returns from first are more that of second. To apply this principle there is a need to understand the following product relationships:

1. **Complementary Enterprise:** Two products are complementary, when increase output of one also results in an increase in the output of the other, with resources held constant.
2. **Supplementary Enterprise:** Two products are supplementary, when an

increase or decrease in the output of one product does not affect the output of the other.

3. **Competitive Enterprise:** Two products are competitive, if output of one product can be increased only through a sacrifice in the output of the other.

The principle of product substitution says that *we should go on increasing the output of a product so long as decrease in the returns from the product being replaced is less than the increase in the returns from the product being added.*

$$\frac{\text{Quantity of output reduced of replaced product}}{\text{Quantity of output increased of added product}} < \frac{\text{Price per unit of added product}}{\text{Price per unit of replaced product}}$$

Decrease in returns < increase in returns

1. Marginal Rate of Product Substitution ($MRPS_{Y1Y2}$):

$$\frac{\Delta Y2}{\Delta Y1}$$

$\Delta Y2$ = quantity of output lost

$\Delta Y1$ = quantity of output gained

2. Inverse Price ratio (PR):

$$\frac{PY1}{PY2}$$

PY2 = price per unit of replaced product

PY1= price per unit of added product

Point where the substitution ratios and price ratios are equal:

$$\frac{\Delta Y2}{\Delta Y1} = \frac{PY1}{PY2}$$

Whether, product Y1 substitutes for Y2 or Y2 substitutes for Y1. The profit rules:

1. If MRPS<PR (substitution ratio is less than inverse price ratio), Profits can be increased by producing more of added product.

$$MRPS_{Y1Y2} = \frac{\Delta Y2}{\Delta Y1} < \frac{PY1}{PY2}\text{ ; increase Y1}$$

2. If MRPS>PR (substitution ratio is more than inverse price ratio), Profits can be increased by producing more of replaced product.

$$MRPS_{Y1Y2} = \frac{\Delta Y2}{\Delta Y1} > \frac{PY1}{PY2} \text{; increase Y2}$$

3. If the MRPS=PR (substitution ratio is equal to inverse price ratio), it is the point of profit maximizing combination.

$$MRPS_{Y1Y2} = \frac{\Delta Y2}{\Delta Y1} = \frac{PY1}{PY2}$$

Table 6.6: Example: Selecting an optimum combination of enterprises.

(P_{Y1} = Rs. 280/quintal; P_{Y2} = Rs. 400/quintal)

Combination	**Y1 (Qtls)**	**Y2 (Qtls)**	**ΔY1**	**ΔY2**	**$MRPS_{Y1,Y2}$**	**PR**	**Decrease in return ($\Delta Y2 \times P_{Y2}$)**	**Increase in return ($\Delta Y1 \times P_{Y1}$)**
A	0	60	-	-	-	-	-	-
B	20	56	20	4	0.20	0.70	1600	5600
C	40	50	20	6	0.30	0.70	2400	5600
D	60	41	20	9	0.45	0.70	3600	5600
E	80	30	20	11	0.55	0.70	4400	5600
F	**100**	**16**	**20**	**14**	**0.70**	**0.70**	**5600**	**5600**
G	120	0	20	16	0.80	0.70	6400	5600

It can be seen from the Table 6.6 that up to E combination MRS is less than PR. But at the F combination MRPS is equal to PR. Therefore, the sixth combination which produces 100 quintals of corn Y1 and 16 quintals of wheat Y2 is the optimum or profit maximizing combination.

6. Principle of Cost or Principle of Minimum Loss

In any business activity the detailed of cost and returns provide an idea about profitability of business. The word cost refers to the expenses on goods and services. Costs are broadly classifying as variable cost and fixed cost. Sum of these costs called total costs. In short-run production activity costs are divided into *fixed and variable cost.*

Different Costs in farm production

1. Fixed Cost (FC): Fixed costs are incurred even in absence of production and it does not vary with the level of output. Fixed costs curve is horizontal straight line parallel to X-axis. The sum of all fixed costs called Total fixed costs (TFC). Fixed cost includes depreciation, rental value of land, interest on fixed capital, land revenue, insurance premium, wages attached to permanent servants, family labour wages, etc.

2. Variable costs (VC): Variable costs vary with the level of output or there is no production no variable cost. The sum of variable costs called Total variable costs (TVC). Graphically TVC has inverse 'S' shape. Variable costs include cost of inputs, customs hiring, wages of casual labour, interest on working capital, electricity charges, etc. Variable costs are very important in decision-making in short-run as the objective of farmer is to recover these costs.

3. Total Costs (TC): Total costs are the sum of Total variable costs and total fixed costs. Its shape is similar to that of total variable costs. Total Costs= TVC+TFC

4. Marginal costs (MC): Marginal cost is the additional cost need to produce one more unit of output. It is the change in the total cost due to the change in output. MC= Change in Total Cost /Change in Output

5. Average Variable Cost (AVC): Average variable cost is the amount spent on the variable inputs to produce a unit of output. Average Variable Cost= Total Variable Cost/ Output

6. Average Fixed Cost (AFC): Average fixed cost is the cost of fixed resources to require for producing one unit of output. Average Fixed Cost= Total Fixed Cost / Output

7. Average Total Cost (ATC): An average total cost refers to the total costs divided by output. Average Total Cost= Total Cost / Output = (TFC+TVC)/ Output.

The farmers sometimes incur loss instead of profit when price of farm produce goes down. Then they should continue the farming with an objective of minimizing losses. Thus, this principle explains, as to how, the producer minimizes losses under adverse price environment.

The farmer should continue the farming till the loss is less than fixed costs. If loss is more than fixed cost then the production should be stopped temporarily, to minimize the loss. In the long run if selling price is less than ATC, continuous losses are incurred. In this situation, the producer should stop the production permanently.

Profit or Decision Rules

a) Short Run:

1. If expected selling price is greater than minimum average total cost (ATC), profit is expected and is maximized by producing where MR = MC.

2. If expected selling price is less than minimum average total cost (ATC) but greater than minimum average variable cost (AVC), a loss is expected but the loss is less than TFC and is minimized by producing where MR = MC.

3. If expected selling price is less than minimum average variable cost (AVC), a loss is expected but can be minimized by not producing anything. The loss will be equal to TFC.

b) Long Run:

1. Production should continue in the long run when the expected selling price is greater than minimum average total cost (ATC).

2. Expected selling price which is less than minimum ATC result in continuous losses. In this case, the fixed assets should be sold and money invested in more profitable alternative.

The following example illustrates the operation of cost principle.

Table 6.7: Example: Costs and returns in groundnut cultivation

Year	**1st**	**2nd**	**3rd**
Groundnut price (Rs./Qtl)	3500	2500	1700
TFC (Rs./ha)	9000	9000	9000
TVC (Rs./ha)	18000	18000	18000
TC (Rs./ha)	27000	27000	27000
Production (Qtl/ha)	10	10	10
Gross Return (Rs./ha)	35000	25000	17000
ATC (Rs./ha)	**2700**	**2700**	**2700**
Net Profit (Rs./ha)	8000	-2000	-10000
Decision whether to cultivate groundnut or not? (On the basis of Minimum Loss Principle)			

Loss if cultivated	No loss	2000	10000
Loss if don't cultivate	9000	9000	9000
Decision	Should be cultivated	Should be cultivated	Should not be cultivated
	Because there is no loss	Because loss is <TFC	Because loss is >TFC
	Price>ATC	Should be permanently stopped if Price<ATC over the years	

7. Principle of Comparative Advantage

A farmer wants to produce those commodities that maximize their net income. He tries to include cropping system as large an area possible of the most profitable crops. The determination of type of farming is based upon the principle of comparative advantage.

The principle states, ***Individuals or regions will tend to specialize in the production of those commodities for which their resources give them a relative or comparative advantage.*** This principle explains *regional specialization* in the production of crops and livestock enterprises. In farm production there are two types of advantage;

1. **Absolute advantage:** The *size of margin* between cost and return from using productive resources represents the absolute advantage. If margin is larger for one farm commodity in one region compared to another, first region has absolute advantage in producing that commodity.

2. **Relative/comparative advantage:** The *return per rupees of investment* from the production represents the relative advantage. This is determined by comparing the two related enterprises present in the region.

Table 6.8: Absolute advantage: Net return per hectare

Crop	Particular	Region A	Region B
Wheat	Total income	18000	13000
	Total expenditure	11000	9000
	Net return per hectare	6000	4000
	Return per rupee	1.63	1.45

Sugarcane	Total income	23000	27000
	Total expenditure	15000	15000
	Net return per hectare	8000	14000
	Return per rupee	1.54	1.80

Table 6.8 shows that, two crops wheat and sugarcane are considered, from the table it can be seen that region A has an absolute advantage in wheat and region B has an absolute advantage in sugarcane because they give net return per hectare of Rs.6000 and Rs.14000, respectively.

Table 6.9: Relative/ Comparative advantage: Return per rupee of investment

Particular	Region A	Region C	
	Groundnut	Groundnut	Wheat
Total income	15000	13500	7000
Total expenditure	8000	8000	5000
Net return per hectare	7000	3500	2000
Return per rupee of investment	1.87	1.68	1.40

Table 6.9 shows that between two crops groundnut and wheat, region A has relative advantage over that wheat of region C because of highest return per rupee of investment (187% of income). But farmers in region C can make more profit by having largest possible area under groundnut relative to that wheat, that income from groundnut is 168 per cent higher than its cost. Thus, the principle of comparative advantage directs a farmer in the selection of those crop and livestock enterprises in the production of which available resources have the greatest relative and not absolute advantage.

8. Time Comparison and Uncertainty Principle

Time is the very important factor in farm decisions. For example, a farmer has to decide between a cereal crop which would be harvested after four months or an orchard which would start giving returns after three years. These decisions involve time and initial capital investment taken by compounding and discounting.

This principle is associated with the investment made on capital assets or fixed assets viz., land, farm buildings, machinery etc. Since, capital expenditure involves cash and benefits over time, it is necessary to adjust for the time value of money which is made through compounding and discounting. Money has time value because money possess earning power, purchasing power which varies inversely with the price level and investment which deals with future and future is uncertain.

Note here: time value of money

1. Compounding cost: The application of this principle involves a process by which the present costs are made to grow with time to make it comparable with future returns.

$$\mathbf{A=P\ (1+r)^n}$$

Where, P is present cost, which grows with the time at compound interest. A is the compound vale of the present cost P invested for 'n' years and 'r' is the rate of the compound interest per rupee.

Example 1: if present cost is 5000 at 10% rate of interest for 4 years, calculate the future value of present investment.

$$\mathbf{A=P(1+r)^n}$$

$=50000 \times (1+10/100)^4$

$=73205$

Example 2: A farmer invested Rs.5000 in establishing the orchard and Rs. 500 each in the beginning of 3 years. At the end of 4 years, he gets a return of Rs. 10000. Find out if this investment of at the rate of 10% per annum compound interest is? The total amount of money which he will have to pay at the end of 5 years:

If $C_1, C_2, C_3, \ldots\ldots\ldots\ldots C_n$ refers to the values paid at the end of the 1st, 2nd, 3rd, ... nth year respectively, the total cost A for the input will be given by:

$A= C_1 (1+r)+C_2(1+r)^2+\ldots\ldots\ldots+C_n(1+r)^n$

$=5000(1+10/100)^4+500(1+10/100)^3+500(1+10/100)^2+500(1+10/100)$

$=9120$

Since at the end of 4 years he has to pay Rs. 9120 and he gets 10000, it is profitable to invest this sum.

2. Discounting Cost: Discounting cost helps to compare the present worth of the future revenue with the present investments. If the present worth of the future revenue is less than the investment, it would be loss to invest the money in resource services now to get the future revenue at the end of a certain number of years.

$$PW = \frac{P}{(1+i)^t}$$

Where, PW is present worth of future money,

P= Money value in future
i= Rate of interest

t=years.

Conclusion

Farm management, includes making and implementing of the decisions involved in organizing and operating a farm for maximum production and profit. Farm management draws on agricultural economics (fall under microeconomics) for information on prices, markets, agricultural policy, and economic institutions such as leasing and credit. Principles of farm management thus provide the farmer to find the practical solution of farm related problems such as production, marketing as well as management of on farm resources. Along with efficient resource management these principles guide to achieve profit maximization level at farm. In this chapter we studied about eight principles of farm management viz., comparative advantage, opportunity cost, theory of cost, product substitution, factor substitution, diminishing marginal return, equi-marginal returns and time comparison. Thus, farm management tools help farmer in solving managerial problems for successful farm business.

References

Bradford, L.A. and Johnson, G.L. (1953). Farm Management Analysis. New York: John Wiley and Sons.

Chaturvedi, D.D. & Gupta, S.L. (2013), Business Economics: Theory and Applications, *International Book House.*

Dean, Joel (1976), Managerial Economics, *Prentice Hall of India.*

Dhondyal, S.P. (1985). Farm Management: An Economic Analysis. *Friends Publications.*

Raju, V.T., and Rao, D.V.S. (2015). Economics of Farm Production and Management. *CBS Publisher & Distributers.*

Reddy, S.S., Ram, P.R., Sastry, T.V.N., and Devi, I.B. (2020). Agricultural Economics. *Oxford & IBH Publishing.*

Website:

1. http://ecoursesonline.iasri.res.in/mod/page/view.php?id=16795
2. https://www.egyankosh.ac.in/bitstream/123456789/88392/2/Block-1.pdf

7

Impact of Agribusiness in Doubling Farmers Income

Yashasvi Ramaul[1*#], Abhay Gautam[2] and Deepak Kumar Dwivedi[3]

[12]MBA Scholar, School of Social Sciences, Central Agricultural University, Imphal, Manipur India

[2]Ph.D Research Scholar, Department of Soil science & Agricultural chemistry, Central Agricultural University, Imphal, Manipur, India

*Corresponding Author Email: yashasviramaul@gmail.com

Abstract

The agribusiness sector plays a vital role in the Indian economy, employing a significant portion of the workforce and contributing to over 10% of the global GDP. However, farmers in India often face challenges such as low incomes and limited market access. To address these issues and double farmers' income, the government has implemented various initiatives, including the Pradhan Mantri Kisan Samman Nidhi (PM-KISAN) scheme. Additionally, agribusinesses are stepping in to alleviate the burden on the government by actively participating in the agricultural value chain. This paper examines key trends in the agribusiness sector, including the rise of new technologies, globalization of the food industry, and the growing importance of sustainability. It highlights the potential of diversified production, emerging opportunities in organic farming and Agri-tourism, and the rapid growth of the fisheries sector. Moreover, the paper explores the role of innovation and technology in enhancing agricultural practices, with a focus on precision agriculture, drones, robotics, artificial intelligence, and blockchain. It emphasizes the need for public-private funding partnerships to promote agricultural research and development, and the importance of access to inputs and services for farmers. Furthermore, the paper discusses market linkage and price stabilization mechanisms facilitated by agribusinesses, such as farmer producer organizations (FPOs), contract farming, e-auction platforms, and commodity exchanges. Lastly, it underscores the significance of developing infrastructure to support the growth of the agribusiness sector. By examining these aspects, this paper provides insights into the challenges and opportunities within the agribusiness sector in India, and the potential for sustainable agricultural growth and improved livelihoods for farmers.

Keywords: Agribusiness, Indian economy, Farmers' income, Market access, New technologies, Sustainability, value & supply chain, Innovation and technology, Market linkage and Price stabilization mechanisms

Introduction

Agriculture is the backbone of the Indian economy, employing over 50% of the country's workforce. However, farmers in India are often poor and have low incomes. In order to double farmers' income, the government has launched a number of initiatives, including the Pradhan Mantri Kisan Samman Nidhi (PM-KISAN) scheme. But agribusinesses are stepping in to alleviate the burden on the government by actively participating.

Agribusiness is the industry that involves the production, processing, distribution, and marketing of agricultural products. It is a broad term that encompasses a wide range of businesses, from farmers to input suppliers to processors to distributors to retailers. The agribusiness sector is a major contributor to the global economy. It accounts for over 10% of global GDP.

The agribusiness sector is important for a number of reasons. It provides food and other products for people around the world. It also creates jobs and generates economic activity.

The agriculture industry is becoming increasingly crucial for economies worldwide. Nations are looking for new and innovative ways to utilize their farmland and reap the benefits. This way of thinking has led businesses within the industry to develop new production, processing, and distribution methods that involve integrating with other companies or farmers to improve overall efficiency.

Rather than trying to accomplish all the activities involved in the agricultural process such as cultivating, equipment use, seeding, and fertilizing alone, companies have become more efficient by employing businesslike practices in their operations. These new developments have transformed the agriculture industry and spurred the agribusiness sector.

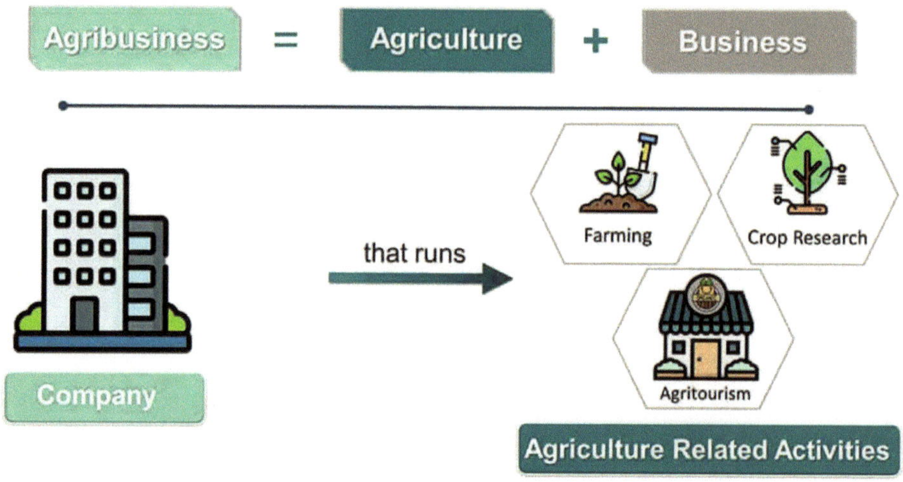

How Agribusiness Works

First step: Production

Agriculture production, where you grow and harvest crops or raise livestock, is the first step in the agribusiness process. This can be done on a local scale by individual farmers or a large scale by agribusiness corporations.

Second step: Processing

The producers must turn harvested livestock and crops into completed commodities. It may involve processing livestock to produce meat products, preserving fruits and vegetables, or milling grains into flour.

Third step: Distribution

The person or business will distribute the products to wholesalers or retailers for sale after processing them. Getting the goods from the processing facility to the point of sale requires transportation and logistics.

Fourth step: Marketing

The companies must then advertise the goods to support and grow The products must subsequently be promoted and brand-awareness developed by the firms. It consists of marketing techniques including advertising, promotions, and others to reach consumers and spur product demand.

Fifth step: Sales

Finally, by selling their goods to customers or other businesses, the businesses will make sales. Contracts and prices may be negotiated, as well as supply chains and customer service.

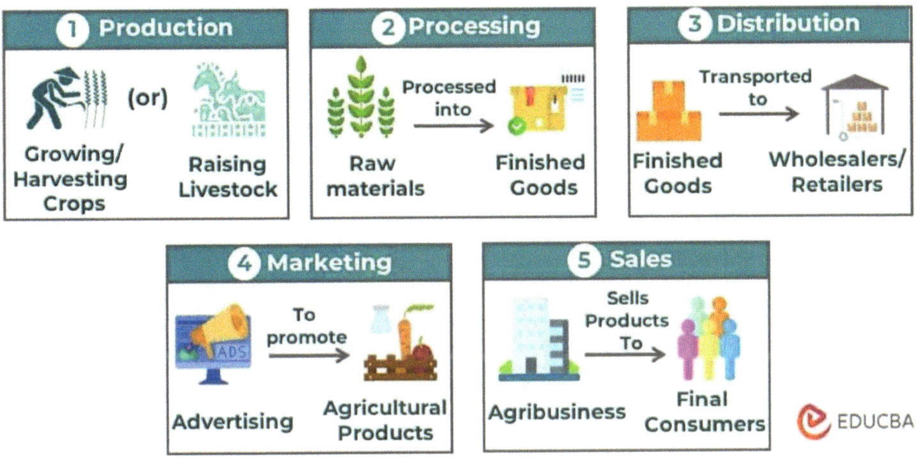

Key Trends in Agribusiness Sector

The rise of new technologies: New technologies, such as precision agriculture and gene editing, are revolutionizing the way agricultural products are produced.

The globalization of the food industry: The food industry is becoming increasingly globalized. This is due to the rise of trade and the increasing demand for food from emerging markets.

Large and growing market: India is the second-largest agricultural producer in the world, with a market size of over $1 trillion. The sector is expected to grow at a CAGR of 3-4% in the coming years, driven by factors such as rising incomes, growing urbanization, and increasing demand for processed food.

Diversified production base: India has a diversified production base, with a wide range of crops and livestock being produced. The country is the world's largest producer of milk, pulses, and spices, and the second-largest producer of rice, wheat, cotton, and sugar.

Emerging opportunities: The agribusiness sector in India is poised for growth, with a number of emerging opportunities in areas as consumers are becoming more aware of the environmental impact of food production. This is leading to a growing demand for sustainable agricultural practices such as natural farming & organic farming.

Organic farming: There is a growing demand for organic food products in India, and the sector is expected to grow at a CAGR of 20% in the coming years.

Agri-tourism: Agri-tourism is a growing trend in India, and the sector is expected to grow at a CAGR of 15% in the coming years.

Fisheries: The fisheries sector in India is growing rapidly, and the country is now the world's second-largest producer of farmed fish.

The growing importance of sustainability: Consumers are becoming more aware of the environmental impact of food production. This is leading to a growing demand for sustainable agricultural practices.

The agribusiness sector is a complex and dynamic industry: It is important to understand the challenges and opportunities facing the sector in order to make informed decisions about the future of food production

Additional Details about Agribusiness Sector

The agribusiness sector is a major contributor to the global economy. It accounts for over 10% of global GDP. The agribusiness sector is made up of a wide range of businesses, from farmers to input suppliers to processors to distributors to retailers. The agribusiness sector is facing a number of challenges, including climate change, increasing demand, and competition.

Climate change: Climate change is affecting agricultural production in a number of ways, such as by causing droughts and floods.

Increasing demand: The global population is growing, and demand for agricultural products is increasing. This is putting pressure on farmers to produce more food.

Competition: The agribusiness sector is becoming increasingly competitive. This is due to the rise of new technologies and the globalization of the food industry.

Despite these challenges, the agribusiness sector is expected to continue to grow in the future.

Role of Agribusiness in Doubling Farmers Income

Agribusiness fosters entrepreneurship and self-sufficiency among farmers. It encourages the formation of cooperatives, farmer producer organizations, and agribusiness clusters, where farmers can collaborate and pool their resources. Through collective marketing, procurement of inputs, and joint investments in infrastructure, farmers gain greater bargaining power, reduced costs, and increased access to markets. These initiatives empower farmers to take control of their own destiny, reducing their reliance on external support systems. Some of these ways are defined below:

Creating Value Chain & Supply Chain

Farmers can establish direct connections with food manufacturers, selling their fresh produce to large processors. A study of farmers in the Krishnagiri cluster, Tamil Nadu, revealed that those linked to the processing industry experienced an average additional income increase of around 49% compared to non-processing industry farmers. The future potential of the food processing market (USD 535 billion) could generate a demand for crops suitable for processing, creating new opportunities for farmers. Food processing companies can engage directly with farmers, offering education, advanced technology knowledge, and cultivating crops for value-added product lines.

India's emphasis on the Green Revolution has led to a concentration of crops like rice, wheat, and sugarcane, resulting in soil degradation, groundwater depletion, and imbalanced market connections. To address these issues and improve farmer incomes, it is crucial to diversify cropping patterns by establishing strong agricultural value chains through Farmers Producers Organizations (FPOs). However, a study found that many FPO members in Telangana, Karnataka, Odisha, Uttar Pradesh, and West Bengal are not actively involved in value chain activities, lacking awareness, infrastructure, and burdened by debt. This situation forces FPOs into distress sales without value addition. While some engage in basic value-added activities, most struggle to invest in assets for significant value addition.

India currently has 15,948 registered FPOs, with 92% being farm-based and only 2.4% exclusively for women. Despite substantial government support, most FPOs struggle to establish sustainable business models and generate significant returns. Encouraging small and marginal farmers, who form the majority, to join FPOs can reduce costs, increase incomes, and ensure food and financial security. Building Agri-value chain infrastructure based on successful models such as milk-based and poultry FPOs, and exploring export markets, can strengthen FPOs and contribute to doubling farmers' income. Access to formal finance and listing on the Social Stock Exchange can mobilize risk capital for FPOs. Support from Cluster-Based Business Organizations for 5 to 7 years can assist FPOs in achieving their goals.

Samunnati (non-deposit taking non-banking financial company), through technology-enabled interventions and collaborative partnerships, provides financial services to over 1,500 FPOs, benefiting 6 million members across 22 states. By improving collective bargaining power in input, credit, and output markets, FPOs can enhance price discovery and improve the lives and livelihoods of India's 126 million small and marginal farmers. In conclusion, it is crucial to strengthen FPOs by ensuring their commercial viability and financial sustainability. This will help overcome challenges in Indian agriculture and have a significant positive impact on farmers' lives and incomes.

Innovation & Technology

The Indian Government aims to double farmers' income by 2022 and has established an Inter-Ministerial Committee to address this goal. The committee recognizes the potential of digital technology, including Artificial Intelligence, Big Data Analytics, Blockchain, and the Internet of Things, to revolutionize agricultural practices in rural India. To support this, various technology interventions have been implemented, such as the Kisan Suvidha mobile application, which provides farmers with critical information on weather, market prices, plant protection, and more. The Indian Council of Agriculture Research has also developed over 100 mobile apps offering valuable information on crops, horticulture, veterinary services, and more. Other initiatives include the mKisan Portal for sending advisories to farmers, the e-National Agriculture Market for online trading, and the Soil Health Card Scheme for soil analysis and nutrient recommendations. Financial assistance and subsidies are provided under different schemes to support farmers in adopting modern farming techniques. Space technology, machine learning, and various dissemination methods are also utilized to enhance agricultural practices. The government's efforts aim to empower farmers with information and resources, ultimately improving their livelihoods and crop productivity.

Whereas agribusiness encompasses a wide range of activities, including production, processing, packaging, distribution, and marketing of agricultural products. By leveraging technology and innovation across these stages, agribusiness has the potential to significantly enhance productivity. Precision farming techniques, such as the use of sensors, drones, and data analytics, enable farmers to optimize resource utilization, improve crop yields, and reduce production costs. Integrated pest management systems and advanced irrigation techniques further contribute to sustainable and efficient agricultural practices.

Some of the ways that innovation and technology are being used in agribusiness:

Precision agriculture: Precision agriculture uses sensors, GPS, and other technologies to collect data on crop health, soil conditions, and other factors. This data can be used to make more informed decisions about crop management, such as when to plant, irrigate, and harvest. Precision agriculture can help farmers to increase yields, improve quality, and reduce costs.

Drones: Drones can be used for a variety of tasks in agriculture, such as crop monitoring, spraying, and fertilization. Drones can help farmers to save time and money, and they can also help to improve the accuracy of crop management decisions.

Robotics: Robots can be used for a variety of tasks in agriculture, such as harvesting, milking, and packing. Robots can help farmers to save time and money, and they can also help to improve the safety of agricultural workers.

Artificial intelligence (AI): AI is being used in agriculture for a variety of tasks, such as crop identification, pest detection, and yield prediction. AI can help farmers to make more informed decisions about crop management, and it can also help to improve the efficiency of agricultural operations.

Blockchain: Blockchain is a distributed ledger technology that is being used in agriculture to track the movement of food from farm to fork. Blockchain can help to improve food traceability, which can help to ensure food safety and quality.

Public-Private Funding Partnership as a Business Model

Public-Private Partnerships (PPPs) are crucial for promoting social and economic development as they combine the efficiency and expertise of the private sector with the long-term perspective and social interests of the public sector. However, in developing countries, the private sector's contribution to agricultural research investment is significantly lower compared to developed countries. For instance, it accounts for only 6.3% in developing countries compared to 55.2% in developed countries (Ramasamy, 2013). To address this disparity and enhance the efficiency of the public research system, there is a pressing need to strengthen PPPs and prioritize evolving technologies to tackle agricultural challenges effectively. Recognizing this need, the International Crops Research Institute for the Semi-Arid Tropics (ICRISAT) established its Agribusiness & Innovation Platform (AIP), which has successfully implemented intervention models empowering Farmers' Producer Organizations (FPOs) and similar collective models. These models have led to increased farmer incomes and rural employment opportunities. One emerging area of focus is the establishment of collection and primary processing centers (PPCs) that streamline value chains, enabling farmers to capture a higher proportion of value.

Access to Inputs & Services

Adani Wilmar Limited is a joint venture between Adani Group and Wilmar International. The company is involved in the production, processing, and marketing of edible oils, rice, pulses, and flour. Adani Wilmar is the largest edible oil company in India and has a market share of over 20%. The company's products are sold under the brand name Fortune.

Godrej Agrovet Limited is a subsidiary of the Godrej Group. The company is involved in the manufacturing and marketing of animal feed, poultry feed, fertilizers, and pesticides. Godrej Agrovet is the second largest animal feed company in India and has a market share of over 15%. The company's products are sold under the brand name Agrovet.

ITC Limited is a diversified conglomerate. The company's agri-business division is involved in the production, processing, and marketing of food products, fibers, and paper. ITC is the largest cigarette company in India and has a market share of over 70%. The company's food products include ITC Sunfeast biscuits, ITC Foods basmati rice, and ITC Farm Fresh milk. ITC's fibers business is involved in the production of cotton, jute, and paper.

Pidilite Industries Limited is a manufacturer of adhesives, sealants, and construction chemicals. The company also has a range of agri-business products, including fertilizers, pesticides, and seeds. Pidilite is the largest adhesives company in India and has a market share of over 60%. The company's agri-business products are sold under the brand name FINOLEX.

Ruchi Soya Industries Limited is a manufacturer of edible oils, pulses, and flour. The company also has a range of agri-business products, including fertilizers, pesticides, and seeds. Ruchi Soya is the third largest edible oil company in India and has a market share of over 10%. The company's products are sold under the brand name Nutrela.

These are just a few examples of the many private firms that are involved in agribusiness in India. These firms play an important role in the Indian economy by providing farmers with the inputs and services they need to produce food and other agricultural products. These agribusiness firms supply high-quality seeds, fertilizers, pesticides, machinery, and other resources necessary for agricultural production. They also offer services such as credit facilities, insurance coverage, technical guidance, and market linkages. This support helps farmers improve their productivity, income, and overall agricultural growth.

Market Linkage and Price Stabilization

Agribusinesses indeed contribute to market linkages and price stabilization for farmers. Here are some real-life examples supported by data:

Farmer Producer Organizations (FPOs) and Agricultural Marketing Companies:

In India, FPOs like Mother Dairy, Amul, and Sahyadri Farms have established strong market linkages for farmers. For instance, Sahyadri Farms has partnered with retail giants like Walmart and Tesco, enabling farmers to directly supply their produce to these chains. This ensures a fair price for farmers and reduces dependence on intermediaries.

As per the data, FPOs have been successful in increasing farmers' income. For example, the National Bank for Agriculture and Rural Development (NABARD) reported that FPOs in India have helped raise farmers' income by 40-60% in certain cases.

Contract Farming:

Agribusinesses often engage in contract farming, where farmers produce crops under predefined agreements with companies. These contracts ensure a ready market for the produce and stabilize prices for the farmers.

An example is PepsiCo India's contract farming model for potatoes. Under this initiative, PepsiCo contracts with potato farmers, providing them with technical support, quality seeds, and assured buyback at predetermined prices. This has helped stabilize potato prices and increase farmers' income.

E-Auction Platforms:

Agribusinesses have developed e-auction platforms to facilitate transparent and competitive bidding for agricultural produce, ensuring fair prices for farmers. The National Agriculture Market (eNAM) in India is one such platform that connects farmers with buyers through online auctions. According to official data, eNAM has witnessed significant participation, with over 1.68 crore (16.8 million) farmers registered and more than 1.89 crore (18.9 million) metric tons of agricultural produce traded on the platform.

Commodity Exchanges:

Agribusinesses participate in commodity exchanges that provide price discovery and risk management mechanisms for agricultural commodities. The Multi Commodity Exchange (MCX) and the National Commodity and Derivatives Exchange (NCDEX) in India are examples of such exchanges. They facilitate futures trading of agricultural commodities, allowing farmers to hedge against price risks. These exchanges contribute to price stabilization and provide farmers with better market access.

These examples illustrate how agribusinesses facilitate the creation of market connections and the stabilization of prices for farmers. By engaging in partnerships, contract farming, e-auction platforms, and commodity exchanges, agribusinesses enable farmers to access equitable markets and receive appropriate compensation for their agricultural products. These initiatives directly contribute to higher income and enhanced economic stability for farmers.

Development of Infrastructure

Agribusinesses in India contribute to infrastructure development, leading to increased income for farmers. Through investments in cold storage facilities, packhouses, irrigation technologies, rural roads, and extension services, agribusinesses play a vital role in enhancing productivity, reducing post-harvest losses, and improving market access for farmers. Agribusiness plays a crucial role in infrastructure development, which ultimately contributes to doubling farmers' income. Here are some real examples from India:

1. Cold Storage and Warehousing Facilities:

» Private agribusiness firms like National Collateral Management Services Ltd (NCMSL) and National Bulk Handling Corporation (NBHC) have invested in the establishment of cold storage and warehousing facilities across India.

» These facilities help farmers store their perishable agricultural produce for longer periods, reducing post-harvest losses and enabling them to sell their products at a later time when market conditions are favorable.

- The availability of such infrastructure supports price stabilization and enhances farmers' income by reducing wastage.

2. Packhouses and Grading Centers:

- Agribusinesses such as Mahindra Agribusiness, ITC Limited, and Super Agri Seeds have set up packhouses and grading centers in India.
- These facilities provide farmers with access to modern sorting, grading, and packaging techniques, ensuring consistency in product quality and meeting market requirements.
- By improving the value of their produce, farmers can command better prices in the market, thereby increasing their income.

3. Irrigation and Water Management:

- Agribusiness firms like Jain Irrigation Systems Ltd and Netafim have played a significant role in promoting efficient irrigation practices in India.
- Through the introduction of drip irrigation, sprinkler systems, and other water-efficient technologies, these companies have helped farmers optimize water usage and increase crop productivity.
- Improved access to reliable irrigation infrastructure allows farmers to cultivate multiple crops, minimize water wastage, and achieve higher yields, ultimately leading to increased income.

4. Farm-to-Market Roads and Supply Chain Infrastructure:

- Agribusinesses, in collaboration with government agencies, have contributed to the development of farm-to-market roads and supply chain infrastructure.
- For example, the National Rural Roads Development Agency (NRRDA) has worked with private agribusiness firms to improve rural road connectivity, enabling farmers to transport their produce efficiently to markets.
- The construction of proper roads and supply chain infrastructure reduces transportation costs, minimizes post-harvest losses, and ensures timely delivery of agricultural products to consumers. This enhances farmers' income by allowing them to reach wider markets.

5. Agricultural Extension Services:

- Agribusiness firms like Mahindra & Mahindra and Tata Chemicals have established agricultural extension services to provide technical assistance and knowledge-sharing platforms to farmers.
- These initiatives involve training programs, demonstrations, and dissemination of best agricultural practices to enhance farmers' skills and knowledge.
- By equipping farmers with updated information and techniques, these services improve productivity, reduce risks, and ultimately contribute to higher income.

Financial Support and Risk Mitigation for Farmers

The Indian government, agribusinesses, and tech startups are working together to provide financial support and risk mitigation to farmers in India.

Government initiatives:

The government has launched a number of initiatives to support farmers financially, including:

- The Pradhan Mantri Fasal Bima Yojana (PMFBY), a crop insurance program that provides financial protection to farmers against crop losses.
- The Kisan Credit Card (KCC) scheme, which provides farmers with affordable credit for agricultural and ancillary activities.
- The Agriculture Infrastructure Fund (AIF), which provides financial support for the development of farm infrastructure, such as cold storage facilities and warehousing.

Agribusiness initiatives:

Agribusinesses are also playing a role in providing financial support and risk mitigation to farmers. They are doing this by:

- Leveraging digital platforms to connect farmers directly with buyers, eliminating middlemen and ensuring better prices for their produce.
- Offering financing options, advisory services, and market intelligence to farmers.
- Collaborating with financial institutions to provide farmers with access to credit.

Tech startup initiatives

Tech startups are also playing a role in providing financial support and risk mitigation to farmers. They are doing this by:

- Developing digital solutions that make it easier for farmers to manage their finances and access credit.
- Providing market intelligence and advisory services to farmers.
- Connecting farmers with buyers and other stakeholders in the agricultural value chain.

The Indian government, agribusinesses, and tech startups are working together to provide financial support and risk mitigation to farmers in India. These initiatives are helping to improve the lives of farmers and ensure the sustainability of the agricultural sector in India.

The government's initiatives are designed to address the challenges faced by farmers, such as low productivity, high input costs, and market volatility.

Agribusinesses are using their expertise and resources to help farmers adopt new technologies and practices, improve their productivity, and access new markets.

Tech startups are using technology to make it easier for farmers to access information, markets, and financial services.

These initiatives are having a positive impact on the lives of farmers in India. They are helping to improve productivity, profitability, and resilience. They are also helping to reduce poverty in rural areas.

Skill Development and Capacity Building

Mahindra Agribusiness: Mahindra Agribusiness, through its Farmer Training Centers (FTCs), conducts skill development programs for farmers across India. These training centers offer practical knowledge on advanced agricultural practices, modern farming techniques, crop management, and post-harvest handling. Farmers are trained on topics such as soil health management, crop protection, efficient irrigation methods, and market linkages. The aim is to empower farmers with the necessary skills and information to improve their productivity, reduce costs, and access better markets. Mahindra Agribusiness has trained over 1 million farmers in India since 2000.

ITC Limited: ITC Limited's Agri Business Division provides skill development programs under the initiative called "Choupal Pradarshan Khet" (CPK). CPK involves setting up demonstration plots in rural areas, where farmers are trained on innovative farming practices and techniques. Farmers receive hands-on training on crop diversification, sustainable farming practices, integrated pest management, and value-added agricultural activities. Through these capacity-building efforts, ITC aims to enhance farmers' income and livelihoods by equipping them with the necessary skills and knowledge. ITC Limited's CPK program has reached over 200,000 farmers since 2010.

Netafim: Netafim, a global leader in drip irrigation solutions, conducts training programs to educate farmers on efficient water management and irrigation practices. These programs focus on teaching farmers about the benefits of drip irrigation, precision agriculture, fertigation techniques, and the proper use of irrigation equipment. By imparting such knowledge, Netafim enables farmers to optimize water resources, increase crop yields, and improve their overall farm productivity. Netafim has trained over 100,000 farmers in India since 2005.

National Skill Development Corporation (NSDC) Initiatives

The NSDC collaborates with various agribusinesses, including private companies and industry associations, to promote skill development in the agricultural sector. They offer programs such as the Pradhan Mantri Kaushal Vikas Yojana (PMKVY), which provides training and certification to farmers in various agricultural skills. Through these initiatives, farmers gain expertise in areas such as organic farming, dairy management, horticulture, livestock rearing, and farm machinery operation. The NSDC has trained over 1 million farmers in India through PMKVY since 2015.

These examples demonstrate how agribusinesses actively engage in skill development and capacity building for farmers in India. By providing training programs and knowledge-sharing platforms, these initiatives equip farmers with the necessary skills and empower them to adopt modern agricultural practices, improve productivity, and ultimately enhance their income and livelihoods.

Need for Government Support and Private Sector Investment

To raise farmers' earnings, the government must provide assistance and the private sector must engage in agribusiness. To raise farmers' earnings, the government must provide assistance and the private sector must engage in agribusiness. The agricultural industry is the backbone of the economy, and the health of the industry directly affects farmers' livelihoods. However, farmers frequently deal with issues including poor pricing, expensive inputs, and restricted market access. These issues can be resolved and farmer incomes can be raised with the help of government assistance and private sector investment.

Government assistance can come in many different forms, including the provision of agricultural input subsidies, funding for agricultural research and development, and the creation of advantageous market circumstances for farmers. Private sector investment can help increase the agriculture sector's production and efficiency while also opening up new markets for farmers' produce.

A Food and Agriculture Organization study shows that Government assistance and private sector investment can boost farmers' earnings by up to 50%, according to research by the United Nations' FAO. The study also discovered that investments from the corporate sector and support from the government can help to lower poverty and food insecurity.

The following are some instances of how governmental assistance and financial support from the private sector can raise farmers' incomes: Governments have the option of offering subsidies for agricultural inputs including fertilizer, seeds, and pesticides. For farmers, this can lower their production costs and boost their income. Investment in agricultural research and development: To create new crop types and farming methods, governments might make investments in agricultural research and development. This may contribute to higher crop yields and better-quality farm goods.

Creating favorable market circumstances for farmers: By making investments in agricultural infrastructure, such as roads, storage facilities, and marketing channels, governments can promote favorable market conditions for farmers. Consequently, farmers may be able to sell their goods for a fair price. The following are some instances of how governmental assistance and financial support from the private sector can raise farmers' incomes: Agricultural input subsidies For agricultural inputs like fertilizer, seeds, and insecticides, governments can offer subsidies. For farmers, this can lower their production costs and boost their income.

Research and development investments in agriculture

Private sector funding the effectiveness and productivity of the agriculture sector can be increased with the aid of private sector investment. This can be accomplished by making investments in modern technology, including irrigation systems and equipment. New markets for farmers' products can also be developed with the aid of private sector investment.

To raise farmers' incomes and strengthen the agricultural sector, government assistance and private sector investments are crucial.

CASE STUDIES

RURAL ENTREPRENEURSHIP THROUGH AMLA FARMING

Mr. Amar Singh from Rajasthan, aged 60 years, initially was receiving an income of INR7 lakh from amla farming. Assured market and remunerative market led him to set up his own processing unit with an initial investment of INR5 lakh that employed both men and women. His unit presently is involved in growing, processing, packaging and transporting. Currently, his turnover is nearly INR26 lakh in one season (August -February).

(Source: National Institute of Agricultural Extension Management)

RIVERINA OILS & BIO ENERGY: INDIA IS THE FASTEST GROWING MARKET FOR CANOLA OIL FROM WAGGA WAGGA

India's growing population means an ever-increasing demand for cooking oil, a major ingredient in Indian food. At the same time, with the incidence of diabetes rising, Indian consumers are becoming more aware of the health effects of the oils they consume. Canola and other vegetable oils are rapidly replacing tallow and palm oils.

Canola oil produced by Riverina Oils and Bio Energy (ROBE) in Wagga Wagga is marketed in India to appeal to this demand. It is labeled as 'pure and chemical free, from pollution free lands' and as an oil to help reduce blood sugar levels, assisting with diabetes control. The high-grade Australian oil is a natural product, processed from non-genetically modified canola seeds and with the high smoke point required for Indian cooking. ROBE was founded by Indian born Australian entrepreneur Dhruv Deepak Saxena, known as DD Saxena. At a custom-built oilseed crushing and refining plant at Wagga Wagga, in New South Wales, it is crushing 25 per cent of the state's canola crop, producing 250 tonnes of refined oil a year.

In 2016 ROBE formed a joint venture with a major Indian food group that has put its canola oil on the shelves of 20,000 shops in India. Should the consumer penetration achieve predicted levels, ROBE will be sending a significant percentage of their oil to India and in time the plant may require additional capacity.

It took DD Saxena half a decade to build the plant and then only a year to reach full capacity. When it was difficult to find finance in Australia, banks in India and other international investors provided capital. Planning approval took longer than expected and the cost of the project doubled, to $150 million. But perseverance has paid off, with DD Saxena hoping to double the plant's capacity in the next few years.

(Source: Department of Foreign Affairs and Trade)

ACIAR: IMPROVING INCOME THROUGH CLIMATE SCIENCES

Agriculture is the biggest employer in India. Climate change is posing a new kind of threat to rural livelihoods. Partnering with Indian researchers and farmers, Australian services and expertise in agriculture and climate science can contribute to increasing productivity and farmer incomes.

Adapting to Climate Change in Asia (2010 - 2015) was a collaborative project between Indian and Australian institutions and funded by the Australian Centre for International Agricultural Research (ACIAR). The project, based in Telangana, tested options for farming communities to enhance their ability to adapt to a changing climate and manage climate risk. The research team worked with policy makers, government departments, NGOs and farmers. Climate Information Centres (CLICs) were one outcome of the project. The CLICs are a platform which combines ICT, traditional and scientific knowledge of weather, and local rainfall measurements to inform farm management decisions. This improves decisions on when to sow crops based on soil moisture and climatic conditions, and increases local capacity to manage climate risk and variability.

Three pilot CLICs were established in the districts of Warangal, Mahboobnagar and Nalgonda. Their success led to subsequent funding and a further 33 centers in Telangana State alone. CLICs will also form an important component of an International Fund for Agricultural Development's Drought Mitigation Project in Andhra Pradesh's five dryland districts.

This partnership improves the ability of smallholder farmers to improve their livelihoods and increase their incomes through resilience to climate variability.

(Source: Department of Foreign Affairs and Trade)

Conclusion

Based on the present resources it is evident that agribusinesses play a crucial role in establishing and strengthening value chains in Indian agriculture. They address challenges related to fragmented supply chains, inadequate infrastructure, and information gaps, by creating efficient networks that connect farmers to markets. By doing so, agribusinesses enable farmers to access inputs, technologies, finance, and markets, ultimately contributing to increasing their income and driving overall growth in the agricultural sector. Infusing fresh life into their centuries-old profession, agribusiness has arisen as a beacon of hope for farmers throughout India, from the sun-kissed plains of Punjab to the verdant landscapes of Kerala. We have seen firsthand the transformative impact of agriculture along our trip, as it serves as a catalyst for doubling farmers' income and opening the door to wealth.

Agribusiness has given Indian farmers access to cutting-edge technology, sophisticated farming methods, and value-added practices, which have completely changed the way they do agriculture. Agribusiness has also encouraged a sense of cooperation and group development among farmers. Small-scale farmers are now able to use their combined strength thanks to the significant support provided by cooperatives and farmer producer organisations. Farmers have more control over their produce and receive fair payments for their labour because of shared marketing channels, infrastructural development, and the elimination of middlemen.

Agribusiness has an impact along the full value chain, not only on the fields. Agricultural products' processing, packaging, and branding have increased their value and opened up new domestic and international market prospects. Farmers have transformed into entrepreneurs that are able to capture a larger portion of the value created by their produce by creating agribusiness hubs and value addition centers. Farmers have strayed from the norm by promoting the growth of high-value crops, horticulture, and floriculture and addressing specialised consumer needs. They have embraced the ideas of agro-tourism, agritourism, and farm stays, giving urban visitors rich experiences while earning extra money.

In terms of policy, the Indian government has acknowledged the enormous potential of agribusiness to improve the lives of farmers. To provide market access, crop insurance, and soil testing services, programmes like the National Agricultural Market (eNAM), Pradhan Mantri Fasal Bima Yojana (PMFBY), and Soil Health Cards have been put into place. These initiatives seek to improve the agribusiness ecosystem and develop favourable conditions for farmers to prosper.

We come to a significant realisation as we draw to a close this investigation of agribusiness and its effect on doubling farmers' income from an Indian perspective. Agribusiness is a potent weapon that empowers farmers, fosters their goals, and transforms their lives; it is not just a trendy term or a transitory fad. We can create a future where farming is no longer linked with hardships and challenges but instead stands tall as a thriving and wealthy sector, boosting the very backbone of our country—the farmers-by using the potential of technology, knowledge, and teamwork.

References

Gulati, A., & Shekhar, S. (2019). The Future of Indian Agriculture: Challenges and Opportunities. *India Policy Foundation.*

Joshi, P. K., & Gulati, A. (2018). Agribusiness in India: Challenges and Opportunities. *Economic and Political Weekly*, 53(45), 41-48.

Pal, S. (2017). The Indian Agribusiness Sector: Growth and Challenges. *Indian Journal of Agricultural Economics*, 72(1), 1-13.

Gulati, A., & Shankar, U. (2018). Agribusiness in India: A Policy Perspective. *Indian Journal of Agricultural Economics*, 73(2), 147-168.

Joshi, P. K., & Gulati, A. (2020). Agribusiness in India: Emerging Trends and Opportunities. *Indian Journal of Agricultural Economics*, 75(1), 1-20

Gulati, A., & Roy, D. (2017). Transforming Indian agriculture: The role of agribusiness. Economic and Political Weekly, 52(16), 32-39.

Mohanty, S., & Sahu, B. (2019). Role of agribusiness in doubling farmers' income: Evidence from India. *Journal of Agribusiness in Developing and Emerging Economies*, 9(4), 505-527.

Mittal, S., & Sharma, S. (2018). Agribusiness development in India: Challenges and strategies. *International Journal of Recent Technology and Engineering*, 8(1), 715-720.

Singh, A., Kumar, R., & Gautam, S. (2020). Agribusiness in India: Current scenario and future prospects. *International Journal of Management, Technology, and Engineering*, 10(1), 17-26.

Ramanathan, R., & Srivastava, L. (2016). Agribusiness and inclusive growth in India: Challenges and opportunities. *South Asian Journal of Socio-Political Studies*, 16(2), 121-135.

Singh, K. R., & Mohanty, R. (2015). Agribusiness development in India: Strategies and challenges. *Indian Journal of Agricultural Economics*, 70(3), 281-293.

Ministry of Agriculture and Farmers' Welfare. (n.d.). Department of Agriculture,

Cooperation & Farmers' Welfare. *Retrieved from https://agricoop.gov.in*

Indian Council of Agricultural Research. (n.d.). ICAR - Indian Council of Agricultural Research. *Retrieved from https://icar.org.in*

My Brooding. (n.d.). How Can India Achieve Its Goal of Doubling the Farmers' Income by 2022? Times of India Readers Blog. *Retrieved from* https://timesofindia.indiatimes.com/readersblog/mybrooding/how-can-india-achieve-its-goal-of-doubling-the-farmers-income-by-2022-33877/

Educba. (n.d.). Agribusiness. Retrieved from https://www.educba.com/agribusiness

Ministry of Agriculture and Farmers' Welfare. (n.d.). Doubling the Farmers' Income. [PowerPoint slides]. Retrieved from https://agricoop.nic.in/sites/default/files/Doubling%20the%20farmers%20income%20.pptx

KPMG. (2021, September). Food processing: Mantra for doubling farmers' income. Retrieved from https://kpmg.com/in/en/home/insights/2021/09/food-processing-mantra-doubling-farmers-income.

Gulati, A., & Sathe, S. (2013). Agribusiness in India: Opportunities and Challenges. *Oxford University Press.*

McKinsey & Company. (2015). the Indian Agribusiness Sector: A Growth Story.

Confederation of Indian Industry (CII). (2017). the State of the Indian Agribusiness Sector.

World Economic Forum. (2016). the Future of Agribusiness in India.

The Hindu Business Line. (2017). The Indian Agribusiness Sector: A Look Ahead.

Economic Times. (2018). India's Agribusiness Sector: A Growth Engine for the Economy.

Kumar, V., & Arora, V. (2018). Digital Farming: The Beacon of New Age Agriculture for Feeding Billions in India. Retrieved from Academia.edu website: https://www.academia.edu/36876121/Digital_Farming_The_Beacon_of_New_Age_Agriculture_for_Feeding_Billions_in_India

Department of Foreign Affairs and Trade. (n.d.). Retrieved from https://www.dfat.gov.au

8

Role of Agribusiness Management in Food Industry

[1]Bhawana Kumari and [2] Dr. Atul Dhingra

[1] Phd Scholar, Department of Business Management CCSHAU, Hisar

[2]Dr. Atul Dhingra Prof. & Head Department of Business Management CCSHAU, Hisar

Abstract

Agribusiness Management not only concerns about maintaining productivity of agriculture related products but also sustainable utilization of resources. It focuses on proper application of inputs (i.e. seed, fertilizers, pesticides, credit etc.), processing of output (i.e. crop grain, milk, meat etc.), food products manufacturing (i.e. bread, cereals, breakfast etc.), and transporting, promoting and selling the food products to the customers and intermediate customers (i.e. whole seller, retailer etc.). The agri-food sector is creating jobs and capable of reducing poverty among society.

Keywords: Sustainable utilisation, processing, transporting and promoting.

Introduction

In 1957, Davis and Goldberg introduced the term 'Agri-business'. Different sectors (i.e. Agriculture input sector, Production sector, Processing manufacturing sector, transport and marketing sector, etc.) are involved in Agri-business management system. So, we can say that Agri-business is related with industry, commerce and trade where industry is associated with the production of materials and commodities while commerce and trade are associated with their promotion and distribution. In current scenario, business has become extremely competitive because of changing customer needs and wants as well as the availability of competitive substitute and cheaper goods in the market. The major objectives of Agribusiness are:

(1) Creating competitive and sustainable private sector which can support the rural development and employment generation activities.

(2) Enhancing productivity of agricultural produce.

(3) Promoting agricultural foods and products as well as agriculture related activities.

(4) Emphasizing more on high potential sectors (such as horticulture, fisheries, livestock, etc.).

The Potential for Agribusiness in India

Agribusiness is important to India for multiple reasons. Such as their contribution to economy, the number of people they employ, strategic reasons of food security, providing raw material to other industries etc. Agribusiness is the largest business in the country significantly larger than other business and growing rapidly. There is a pressing need to focus on agriculture and grow agribusiness.

Agri-business basically involves three sectors:

1. Input sector: It deals with the supply of inputs, both consumables, and durable products, required by the farmers for producing crops, raising livestock, and other allied enterprises. These include seeds, fertilizers, plant protection chemicals, machinery, and fuel.
2. Farm sector: This sector undertakes the production of crops, livestock, fish, and forest products.
3. Product marketing and services sector: This includes firms or industries that are involved in processing (product transformation), storage and warehousing, transportation, wholesaling and retailing of products, and marketing of the finished products to meet the dynamic needs of consumers.

Scope of Agribusiness

1. All the business activities which ensure the availability of food and fiber products at the desired place, in the required form and quantity, at the right time come within the purview of agribusiness. In other words, all business activities in the input supply sector, farm production sector, and food processing sector make up the subject matter of agribusiness. In addition to these, logistics and marketing

of food and fiber products also constitute activities within the spectrum of agribusiness activities.

2. In India, varied agro-climatic conditions ranging from sub-tropical to tropical and temperate, and farming environments in terms of soil types, allow the cultivation of varied types of agricultural commodities and the rearing of different types of animals for food and fiber.

3. Consumer needs and preferences for food commodities, especially processed products, are changing rapidly in India. The dynamic agribusiness sector is continuously striving to meet current consumer preferences and satisfy demand in both domestic and global markets.

4. This has created a growing demand for different types of agricultural inputs such as improved seeds, feed and fodder, inorganic fertilizers, bio-fertilizers, and plant protection chemicals

5. Diverse types of commercial enterprises including manufacturing and production, financial and technical services, processing, and marketing, function in the agribusiness sector. These diverse agribusiness enterprises offer the scope for employment and utilization of a heterogeneous combination of labour, materials, capital, and technology.

The Components of Agribusiness

The Agribusiness System thus can be said to comprise the following six sub-systems:

(i) Government policies and programmes. (ii) Research and extension programmes of the Government (iii) Farm inputs suppliers or input sector (iv) Agricultural production sector (v) Processing – manufacturing or agro-processing sector, and (vi) The marketing of agricultural products –distribution sector.

(i) Government Policies and Programmes: Government policies and programmes relating to the raising of crops and other enterprises such as livestock, poultry, fisheries, etc., form one component of the agribusiness system in the country.

(ii) Research and Extension Programmes of Government: Agricultural research network comprises the central and state agricultural universities and research centers and institutes under the Indian Council of Agricultural Research (ICAR) This vast network of research institutions in India is devoted to developing new knowledge/ technology on crop production and post-harvest management of agricultural produce.

In addition to these, there are also industrial research and development institutions including technical colleges and institutes which are engaged in developing new production and processing technology, machines, equipment, etc. In order to disseminate the information and technologies developed by the research and development organizations, the Government has established an extension system.

(iii) Farm Inputs Suppliers or Input Sector: This sector produces and supplies farm inputs such as fertilizers, plant protection chemicals, seeds, feed, and farm machines used by the production sector of agribusiness. The input firms are a major part of the agribusiness sector and produce proximately 75 percent of all the inputs currently used in production. The growth in the agriculture input sector exceeds the growth in other sectors of the agribusiness system in most developing countries. The reason for this is obvious – increasing demand and dependence on purchased inputs of all types.

(iv) The Agricultural Production Sector: The production sector is at the center of much of the developments in agribusiness. The production efficiency on farms is growing as farmers are increasingly buying yield-increasing farm inputs from the market. Agriculture is fast becoming commercialized and farm production is getting increasingly transferred to the manufacturing-processing sector and sold in domestic as well as global markets. A new trend of diversification from high volume - low value to low volume-high value crops is being experienced in countries like India.

(v) The Agro Processing and Manufacturing Sector: The processing or manufacturing sector includes all the individuals and firms that process raw agricultural commodities to produce value-added products like converting wheat into flour or raw milk into pasteurized milk or flour into bread. There are thousands of processed food products that are demanded by consumers and this sector manufactures them for the final consumers. The size and attractiveness of this sector are evident from the fact that big multinational companies like Pepsi, Coca-Cola, Hindustan Unilever, Kellogs, and Nestle, are involved in food manufacture.

(vi) The Distribution - Marketing Sector: The distribution-marketing sector plays a major role at each level in the agribusiness supply chain from input supply to production to processing and consumption. Millions of people and firms are engaged in this sector. On one hand, the sector makes available farm inputs to the production sector, on the other hand, this sector makes available processed products to the final consumers. The packaging industry, transport companies, warehouses, advertising companies, insurance companies, wholesale firms, retail outlets, etc. are all part of this sector.

Agribusiness, value chains and the rapid transformation of agrifood systems:

Agrifood systems are currently undergoing a rapid transformation driven by urbanization, globalization, diet diversification, concentration, and expansion of food markets and trade, among other underlying trends. Coping with these changes requires a broader systems perspective that emphasizes agrifood value chain coordination, value creation and the institutional setting in which value chains operate. This cannot be done without embracing agribusiness development.

The agribusiness sector, which comprises the business activities performed from farm to fork, is a major generator of employment and income worldwide and contributes to food security and nutrition. It covers the entire value chain, including the supply of agricultural inputs, the production and transformation of agricultural products, and their distribution to final consumers. The relevance of agribusinesses is underlined by the fact that they

» Are often the main source of off-farm employment in rural areas of poor countries;

» Have positive effects on poverty reduction and women's empowerment in countries where high-value agrifood exports are produced;

» Create off-farm employment opportunities in agro-industrial firms located in rural areas, improving the income of rural households through wage employment and spillover effects that can increase on-farm agricultural productivity through greater liquidity to purchase inputs and increased capacity to adopt technologies.

» Help to forge the necessary link between the agriculture and manufacturing sectors, which in turn can catalyse the development of broader manufacturing industries by providing material inputs for food processing, textiles and biofuels.

Seen as an engine for growth, agribusiness and its related industries are receiving increased attention in policies and strategies that aim to promote investments in agro-enterprises and agrobased value chain development. This has prompted a need for deeper understanding of the elements that form a conducive business climate, or enabling environment, for the development of agribusinesses. The agribusiness and sustainable food value chain (SFVC) work involves the collection and analysis of information and experiences in order to understand changing realities and trends. Based on this, support is provided to member countries' governments by advising them on policies and strategies to improve agribusiness competitiveness and food value chain and food system performance, including fostering greater public-private collaboration and linkages among business partners. It also produces training materials, in particular for policy-makers, development practitioners and small and medium agricultural enterprises (SMAEs).

FAO role in agribusiness and value chain development

In support of agribusiness and food value chains, FAO focuses on enabling environment aspects and on building the capacities of ministries of agriculture to effectively engage with private sector partners, especially on the following public objectives: inclusion of farmers and SMAEs, investment promotion and increased efficiency. FAO agribusiness work includes the development and dissemination of a series of tools and policies used to stimulate agrifood industry and food value chain development and support the inclusion of smallholder farmers and small rural food enterprises in rapidly transforming agrifood systems in developing countries. The main policy issues addressed include:

» Responsible contract farming for inclusive market access providing capacity building, technical and legal support to field projects, advocacy, and dissemination of knowledge.

» Developing sustainable food value chains (SFVCs) as a holistic approach to structural transformation, integrating multidimensional value-added and sustainability concepts and applying them in the measurement, analysis and improvement of food system performance.

» Public-private partnerships or public-private-producer partnerships, which are innovative partnerships that bring together producers, agribusiness, government and civil society actors. They are increasingly being promoted as a mechanism for pooling much-needed financing while mitigating some of the risks of doing business in the agriculture sector.

» Fostering a territorial approach as a means to attract transformative investments in agribusiness and coordinate cross-sectoral objectives. Particular attention is paid to agro-industrial parks and agrobased clusters, economic corridors, and Special Economic Zones.

» Encouraging small and medium enterprise development, entrepreneurship, agribusiness start-ups and innovation through inclusive business models and agribusiness incubators.

» Institutional procurement programmes, by which governments award public tenders not only to respond to the state's procurement needs but also to simultaneously address social, environmental or economic concerns. FAO collaborates on this topic with external institutions, such as the International Food Policy Research Institute (IFPRI), International Fund for Agricultural Development (IFAD), Inter-American Development Bank (IDB), Organisation for Economic Co-operation and Development (OECD), International Institute for the Unification of Private Law (UNIDROIT) and the World Bank.

Indian agribusiness ecosystem is significantly fragmented and unorganized across the value chain:

The agribusiness ecosystem comprises the business activities performed from farm to fork, covering the entire value chain, from the supply of agricultural inputs, the production and transformation of agricultural products, and their distribution to final consumers. Driven by rapid urbanization, diet diversification, evolving consumer preferences and expansion of food markets, the agribusiness ecosystem has been expanding to segments such as e-commerce and hyperlocal.

Agriculture contributes to 16% of India's GDP:

Agriculture plays a vital role in the Indian economy, contributing ~16% (US$2,842 billion in 2018 at constant 2010 US$) to the gross domestic product (GDP) and providing employment to 43% of the total workforce. India ranks second after China, accounting for 11.9% of the global agriculture Gross Value Added (GVA) of US$3,320.4 billion. Agriculture is also an important part of internal and external trade, positioning India as a significant agri-exporter, with the sector contributing 12% to India's exports. The sector also impacts non-agricultural segments such as consumer products, retail, chemicals and e-commerce that are dependent on agricultural cash crops for raw material (tea, coffee, cotton, jute, sugarcane, oilseeds, etc.)

The agriculture and allied sector's GVA at constant 2011-12 prices grew at a CAGR of 3.1% during FY12-19. The contribution of agriculture to the GVA is on a steady decline, decreasing from 18.2% in 2014-15 to 16.5% in 2019-20. The share has been declining on account of relatively higher growth performance of non-agricultural sectors. Indian agriculture broadly comprises of farming (crops and horticulture) and forestry, livestock (milk, eggs, meat) and fisheries.

Numerous technologies are being leveraged by agritech players to drive efficiencies:

- » Data analytics and machine learning
- » Data and platforms for price transparency
- » Imaging and AI to monitor crop quality
- » Platforms for produce traceability
- » Robotics and drones for cultivation/ harvesting

Evolving consumer behaviour and supply chain efficiencies are key demand drivers for agritech

Innovations in the agriculture sector are changing the way food has been grown and distributed. These innovations were triggered by factors including climate change, reduction in water availability, continuous increase in tolerance of pests to crop protection measures, decrease in quality of soil due to unsustainable farming and labor shortage.

Increasing urban population and a move towards healthier food patterns is driving a change in consumer dynamics

Supply chain improvements and efficiencies can help lower food wastage and improve productivity for farmers

Climate-resilient technology and data-driven weather forecasts can help combat climate change and drive efficiencies.

The agri-food sector: creating jobs and capable of reducing poverty

In this issue, we will use the term agribusiness in a broad sense. It includes processing agricultural commodities, packaging and distributing them as marketable products as well as the related services (supplying fertiliser, seed and equipment) and, by extension, all agricultural production systems. While the agri-food sector sensu stricto only includes food processing and supplying inputs, in practice, the contractual, institutional, functional and dependency ties between processing and production mean that we cannot really think of them separately and they are very often integrated (Barrett et al., 2001).

In its 2008 World development report, the World Bank insisted on the need to invest in agriculture and agribusiness to boost growth and reduce poverty. Ten years later, a number of researchers are revisiting these findings and emphasising the role of agribusiness in economic development.

Agribusiness, environment, equity: The pitfalls to be avoided

The reason why agri-food is often seen as the antithesis of sustainable development is because of its perceived negative externalities both for the environment (Barrett et al., 2001) and for small producers (Reardon et al., 2009).

Revisiting the last few decades of the twentieth century, Reardon, Barrett et al recall that deregulated globalisation, lower tariff barriers and transport costs, price wars and technical progress have all shaped agro-industrialisation, i.e., the tendency toward vertical integration, the scramble for higher yields, use of intensive chemical inputs and downward pressure on producers' prices, etc. While the externalities from this wave of agro-industrialisation are not systematically negative, when they are, the price paid tends to be very high both for the environment (deforestation, GHG emissions, biodiversity loss, soil depletion and pollution of aquatic environments) and in terms of equity (exclusion of smallholders from the market or even from their own land if traditional land rights are not protected, inequitable distribution of value within channels controlled by dominant firms (BASIC 2014), poor working conditions, etc.).

On the other hand, technical progress makes it possible to develop crop varieties that consume less water and international export outlets can help secure long-term agricultural employment and avoid land artificialisation (Barrett et al., 2009). In reality, the conditions under which agro-industrialisation takes place play a determining role: national regulations can provide a framework for preventing, reducing or offsetting negative externalities insofar as its development is actually being controlled – which is not necessarily the case in countries with insufficient technical, financial and human resources. Voluntary CSR-type initiatives or certified standards can also be favourable factors. Lastly, stronger demand for higher-quality, eco-friendly products and fairer trading practices can enhance environmental and social performances.

In other words, agro-industrialisation that complies with sustainable development principles is possible if it manages to include small producers, respect the environment and remain competitive from both a price and a quality perspective, even in the face of fierce international competition. It will primarily focus on providing at least some answers to these questions which are crucial to several of the UN's sustainable development goals (SDGs) (i.e., hunger, poverty, employment and sustainable production and consumption methods).

Challenges faced by agribusiness industries

The global food system is expected to provide safe and nutritious food to a population that will likely grow from 7.5 billion people today, to nearly 10 billion by 2050. Not only will there be more mouths to feed, but as incomes grow in emerging and developing economies, so too will the demand for meat, fish, and dairy.

However, food production is only one aspect of the food system. The agro-food sector also provides a livelihood for millions of people. Globally, most of the people living in extreme poverty are in rural areas where food production is often the most important economic activity. There are an estimated 570 million farms worldwide today, and millions of other people work in food-related jobs.

The global food system also has a large environmental footprint. In fact, agriculture occupies nearly 40% of the earth's surface, far more than any other human activity. In addition, irrigation of agricultural crops comprises 70% of global water use, and agriculture directly contributes to around 11% of global greenhouse gas (GHG) emissions (mostly through cattle). Expanding agricultural land can also lead to deforestation, additional GHG emissions, and a loss of biodiversity.

These three challenges – feeding a growing population, providing a livelihood for farmers, and protecting the environment – must be tackled together if we are to make sustainable progress in any of them. But making progress on this "triple challenge" is difficult, as initiatives in one domain can have unintended consequences in another.

Sometimes, the consequences are positive. For instance, raising farm productivity can generate income growth in agriculture, make more food available for consumers at lower prices, and – in some cases – reduce pressure on the environment. But sometimes the consequences are negative and require balancing trade-offs. For example, policies to increase the environmental sustainability of agriculture could impose increased costs on farmers and lead to higher prices for consumers.

In other words, policies that address one part of the triple challenge often end up creating synergies (positive effects) or trade-offs (negative effects) with respect to other objectives—and a single-issue perspective on any objective can lead to unintended impacts on other objectives. Competing objectives and complex interactions, along with multiple stakeholders with a range of concerns, should make us cautious when specific ideas are proposed as "silver bullets" to fix the food system.

So what can policy makers do to address these important challenges, taking into account their interconnectedness? How should they find out if and when there is a conflict between two or more objectives? How should they deal with stakeholders who may resist an initiative they fear could harm their interests? And how should they co-ordinate with policy makers in other agencies or ministries, and with counterparts in other countries?

To begin the process of answering these difficult questions, the OECD organised a Global Forum on Agriculture in May 2019 to exchange ideas about the most important challenges facing the global food system today (the triple challenge), and the obstacles that stand in the way of overcoming them. Importantly, the conversation included views from a range of stakeholders affected by agro-food policy decisions – including farmers, traders, food manufacturers, consumer representatives, agricultural input suppliers, researchers, environmental NGOs, and policy makers. The OECD will build on this discussion to assess the main obstacles to achieving better policies for the global food system, and to identify good practices to help overcome them.

Future policies

Just like a good meal is a balanced meal, good policies will need to strike a balance between the different objectives of the triple challenge facing the global food system today. And just like a good meal depends not only on the chef, but also on the quality of the ingredients – so too will good policies depend not only on the policy maker, but also on the input from many stakeholders. Given the scale and complexity of these challenges, policy makers may need to experiment with new recipes to cook up a set of policy solutions that are to everyone's taste.

Conclusion

Agribusiness play an important role in food supply. It deals with input, farm or production marketing. So that inputs are supplied in field and output are marketed in the markets. Value chain and supply chain also adds some features to it. Government also provides some policies to improve food security. Agribusiness also provides 16% to India GDP and also provides employment. There are also some challenges which are based by agribusiness like providing food to increasing population, provide livelihood to farmers and to maintain environment security.

References

B.Jonathan, D.Koen &G.Celine(2019) the three key challenges facing agriculture and how to start solving them.

B.Rajeev &J.Ivi(2019) Sustainable food supply chains. Planning design and control through interdisciplinaries methodologies, 23-27.

Agritech –towards transforming Indian agriculture(2020).

Food and agriculture organisations of united nations.

9

Technology Adoption in Agribusiness

Shaik Reshma Sulthana[1] and Remalli Anthony[2]

[1]Ph.D. Scholar at IABMI, AAU, Anand, Gujrat-388110
[2]Ph.D. Agricultural Economics, Department of Agricultural Economics, Anand Agricultural University, Anand, Gujrat-388110

Abstract

The adoption of technology in agribusiness has gained significant attention in recent years as a means to address the challenges faced by the agriculture industry. It encompasses an overview of the key factors driving technology adoption in agribusiness and highlights the benefits and implications of implementing technology solutions in this sector. These technologies enable farmers and agribusinesses to enhance productivity, optimize resource allocation, and improve overall efficiency. By utilizing sensors, drones, and satellite imagery, precision agriculture allows for real-time monitoring of crop health, soil conditions, and water usage, leading to informed decision-making and targeted interventions.

In conclusion, the adoption of technology in agribusiness holds immense potential for transforming the sector by improving productivity, resource efficiency, and supply chain management. However, addressing challenges related to access, affordability, and knowledge gaps is crucial for ensuring inclusive and sustainable technology adoption. Policymakers, industry stakeholders, and technology providers must collaborate to create an enabling environment that promotes the adoption and integration of technology solutions in agribusiness, ultimately driving agricultural development and contributing to global food security.

Keywords: Smart Agribusiness, Digital technologies, Supply chain Management

Introduction

Agriculture continues to play a significant role in the creation of income and provide food for a large number of people worldwide. It is changing to more sustainable methods all across the world, the shift is the result of customers' rising environmental consciousness and demand for more environmentally friendly food. The agribusiness industry must adapt new procedures while maintaining its bond with the earth in order to feed the globe and develop crucial environmental measures. Despite having roots that go all the way back to the beginning of civilization. Modern agriculture firms are looking for innovative strategies, business models, and procedures to address new agricultural trends that are transforming the sector.

Smart Agribusiness

The term "**smart agribusiness**" refers to the use of digital technologies to boost the productivity of all parties involved in interconnected and interdependent value chains in agriculture. The main objective of smart agribusiness is to employ digital technologies to increase profit for all stakeholders along the value chain while ensuring the sustainability of the economy, environment, and natural resources. Crops, livestock (and other animal-based products like honey and milk), fisheries, and non-food items are the four areas of agriculture where Smart Agribusiness finds applications (such as leather and textile).

A trend called "smart farming" places a focus on using information and communication technologies in the cycle of managing a cyber-physical farm. More robotics and artificial intelligence will likely be used in farming as a result of new technologies like the Internet of Things and cloud computing.

Advanced Technologies for Smart Agribusiness

Digital technology, which is a component of the so-called "Fourth Industrial Revolution (4IR)," is what allows for smart agribusiness. Digital technologies have expanded quickly as a result of advancements in hardware scalability, Internet of Things (IoT), greater internet connectivity, computing and data storage performance, accessibility, and cost. Digital technologies have the potential to boost innovation, reduce costs associated with solution scaling, increase transparency, and support an informed, evidence-based transformation of the agriculture industry.

Blockchain

Blockchains are decentralized, secure ledgers. According to Olsen and Tomlin's (2019) definition, it is distributed because its data is kept on a peer-to-peer network and may be written and read by several (potentially authorized) organizations. Additionally, it is secure because once a "block"—i.e., information written by an entity—has been added to the chain, it cannot be changed unilaterally.

Internet of Things (IoT)

The term "Internet of Things" describes a network of gadgets (such as sensors) that can connect to the internet and interact with one another, as well as be watched and managed from a distance. In recent years, the adoption rate of the Internet of Things has increased tremendously due to the huge cost and processing speed reductions for these sensors as well as the major advancements in measurement and communication technology.

Precision Farming

Precision Framing is the science of improving crop yields and assisting management decisions using high-technology sensor and analysis tools. It is a new concept adopted throughout the world to increase production, reduce labour time, and ensure the effective management of fertilizers and irrigation processes.

Artificial Intelligence (AI)

Unlike the "natural intelligence" that people possess, AI imitates natural intelligence by utilizing computers to analyze external data, learn from it, and then use what it has learned to undertake descriptive, predictive, or prescriptive analysis.

Drones

Drone use has become an essential part of large-scale precision farming operations already. The data collected from drone recording fields help farmers plan their planting and treatments to achieve the best possible yields.

Robotics

The objective of **agricultural robotics** is to help the sector in its efficiency and in the profitability of the processes. In other words, mobile robotics works in the agricultural sector to improve productivity, specialization, and environmental sustainability. Labour shortages, increased consumer demand, and high production costs are some of the factors that have accelerated automation, with the aim of reducing costs and optimizing harvests.

Sensors

Sensors used in smart farming are known as agriculture sensors. These sensors provide data that assist farmers to monitor and optimize crops by adapting to changes in environmental conditions. They are installed on weather stations, drones, and robots used in the agriculture industry.

Cloud Computing

Cloud computing is one of the most dynamic and popular computing models that combines the concept of cluster and grid computing. This means it offers a single interface for users to perform intensive tasks by utilizing similar computer resources from multiple locations. The framework of cloud computing provides a shared pool of IT resources for processing, networking, and storage. Users can apply to use these resources as per their requirement and only pay for the same. This helps them in acquiring IT capabilities without spending much on the deployment and maintenance of IT assets. The technology also enables them to scale up or down their spending on the given resources based on their business requirements at any given time. Cloud Computing in agriculture, though a comparatively new concept, allows farmers easy access to information and, in return, helps them create easy access to their produce.

Big Data

Big data primarily refers to data sets that are too large or complex to be dealt with by traditional data-processing application software. Big data provides farmers with granular data on rainfall patterns, water cycles, fertilizer requirements, and more. This enables them to make smart decisions, such as what crops to plant for better profitability and when to harvest. The right decisions ultimately improve farm yields.

Farm Machines

One of the challenges that farmers face nowadays is the need to satisfy labour. There is an increasing cost of labor, which calls for better approaches to ensure less cost of labour. The introduction of combined harvesters and planters simplifies the process. Production and time are some of the important elements in agriculture. It is important, therefore, to plant early, harvest in time, as well as ensure that the yield is stored within the right time. The use of modern technology in agriculture ensures that farmers grow vast food within the shortest time possible.

GPS technology has been used in the development of autopilot sprayers and tractors that do not require any driver. Such technology is important in agriculture in that it promotes better and more efficient farming practices. For example, autopilot tractors and sprayers are equipped with tracking systems that eliminate human error and in the end save on fuel and equipment.

Use of GPS in fields documentation

GPS is becoming a common technology in agriculture. For example, modern agriculture involves the use of GPS to document the status of the farmland. Through the GPS, it is easy to determine and document the yields from a given farm, as well as record the application rates. Such technologies are useful in that the farmers can rely on the collected and recorded data for reference when making any decisions. The recommendable documentation technology is the yield map, which can be used to offer a summary of entire year's activities. Such maps are highly useful as they can give a wide range of information about just anything such as the status of the drainage system in your field.

Key factors driving adoption of technology in Agribusiness

The adoption of technology in agribusiness is driven by several key factors. These factors include:

1. **Increasing global food demand:** The world's population continues to grow, leading to an increased demand for food. Technology adoption allows agribusinesses to meet this demand by improving productivity and efficiency in agricultural practices.

2. **Need for improved productivity:** Technology offers tools and solutions that help farmers enhance productivity by optimizing resource allocation, minimizing waste, and maximizing output. Automated machinery, precision agriculture techniques, and data-driven decision-making enable farmers to achieve higher yields with fewer resources.

3. **Labour shortages and rising labour costs:** Many regions face labour shortages in the agricultural sector, and the cost of labour is rising in various parts of the world. Technology adoption, such as robotics and automation, can help alleviate these challenges by reducing the dependency on manual labour and increasing operational efficiency.

4. **Climate change and resource constraints:** Climate change poses significant challenges to agriculture, including unpredictable weather patterns, water scarcity,

and soil degradation. Technology adoption enables farmers to monitor and adapt to changing conditions, implement sustainable practices, and optimize resource usage, thus mitigating the impact of climate change on agricultural production.

5. **Advancements in digital connectivity:** Improved access to affordable and reliable internet connectivity has created opportunities for technology adoption in agribusiness. Internet of Things (IoT) devices, cloud computing, and data analytics can be leveraged to collect and analyze real-time data, enabling farmers to make data-driven decisions and optimize their operations.
6. **Increasing availability and affordability of technology:** Technological advancements have led to the development of cost-effective and user-friendly solutions for agribusinesses. From low-cost sensors and drones to mobile applications and farm management software, the availability and affordability of technology have made it more accessible to farmers of all scales.
7. **Market demands and consumer preferences:** Consumers are increasingly conscious about the origin, quality, and sustainability of the food they consume. Technology adoption allows agribusinesses to meet these market demands by improving traceability, ensuring food safety, and implementing sustainable agricultural practices.
8. **Government support and policies:** Governments play a crucial role in promoting technology adoption in agribusiness through supportive policies, incentives, and funding. Policy frameworks that encourage research and development, infrastructure development, and knowledge transfer facilitate the adoption and integration of technology solutions in the agriculture sector.

These factors collectively drive the adoption of technology in agribusiness, offering opportunities for increased productivity, efficiency, and sustainability in the agricultural sector.

Benefits of implementing technology adoption in Agribusiness

Implementing technology adoption in agribusiness brings numerous benefits to farmers, agribusinesses, and the agriculture sector as a whole. Some key benefits include:

1. **Enhanced productivity:** Technology enables farmers to optimize their operations, leading to increased productivity. Precision agriculture tools, such as sensors, drones, and satellite imagery, provide real-time data on soil conditions, crop

health, and water usage, allowing farmers to make informed decisions and take targeted actions. This results in higher crop yields and improved efficiency in resource allocation.

2. **Improved resource management:** Technology adoption facilitates efficient management of resources, such as water, fertilizers, and pesticides. By utilizing data-driven insights, farmers can optimize the application of these resources, reducing waste and environmental impact. This not only conserves valuable resources but also contributes to sustainable agricultural practices.

3. **Risk mitigation:** Technology solutions assist in mitigating risks associated with weather fluctuations, pests, and diseases. Early detection systems, weather forecasting tools, and pest monitoring devices enable farmers to take proactive measures, minimizing crop losses and increasing resilience. By integrating technology, farmers can better adapt to climate change and address potential risks in a timely manner.

4. **Streamlined supply chain management:** Technology adoption improves supply chain management in agribusiness. Blockchain technology, for example, enhances transparency and traceability by creating an immutable record of transactions and movements along the supply chain. This ensures food safety, quality control, and efficient recall management. Additionally, smart logistics and inventory management systems optimize storage, transportation, and distribution, reducing post-harvest losses and ensuring timely delivery to the market.

5. **Access to market information:** Technology provides farmers with access to real-time market information, enabling them to make informed decisions about crop selection, pricing, and market timing. Online platforms and mobile applications connect farmers with buyers, facilitating direct market access and reducing dependence on intermediaries. This empowers farmers to negotiate fair prices and expand their market reach.

6. **Data-driven decision-making:** Technology adoption generates vast amounts of data that can be analyzed to gain valuable insights. Data analytics and predictive modeling enable farmers to make data-driven decisions, improving operational efficiency and profitability. By understanding patterns and trends, farmers can optimize their practices, manage risks, and identify opportunities for growth.

7. **Adoption of sustainable practices:** Technology supports the adoption of sustainable agricultural practices. From precision farming techniques that reduce chemical inputs to smart irrigation systems that conserve water, technology enables

farmers to implement environmentally friendly methods. This contributes to the preservation of natural resources, biodiversity, and ecosystem health.

8. **Empowering small-scale farmers:** Technology adoption can benefit small-scale farmers by providing them with access to knowledge, information, and market opportunities that were previously limited. Mobile applications, remote sensing, and e-learning platforms enable small-scale farmers to access agricultural expertise, receive timely advice, and gain market insights, leveling the playing field and improving their livelihoods.

These benefits contribute to a more efficient, sustainable, and resilient agriculture sector.

Impact of Technology Adoption on supply chain management of Agribusiness

Technology adoption plays a vital role in improving supply chain management in agribusiness. Here are some ways in which technology enhances supply chain management in the agriculture sector:

1. **Enhanced traceability:** Technology, such as blockchain, enables improved traceability throughout the supply chain. By creating an immutable and transparent record of transactions and movements, it becomes easier to track the origin, production practices, and handling of agricultural products. This enhances food safety, quality control, and regulatory compliance. Consumers, retailers, and other stakeholders can have confidence in the authenticity and integrity of the products they buy.

2. **Real-time monitoring and visibility:** Technology allows for real-time monitoring and visibility of key supply chain parameters. Sensors, RFID tags, and loT devices can be used to track the location, temperature, humidity, and other relevant factors during the transportation and storage of agricultural products. This data helps identify any deviations from desired conditions, enabling prompt corrective actions and reducing product spoilage or losses.

3. **Efficient inventory management:** Technology adoption enables more efficient inventory management in agribusiness. Automated systems and software applications can track inventory levels, monitor expiration dates, and manage stock replenishment. This ensures that the right amount of products is available at the right time, minimizing wastage and reducing the risk of stockouts or overstocking.

4. **Optimized logistics and transportation:** Technology facilitates optimized logistics and transportation in the agricultural supply chain. Route planning and optimization algorithms can minimize transportation costs, fuel consumption, and carbon emissions. GPS tracking systems enable real-time monitoring of delivery vehicles, ensuring timely and efficient transportation of agricultural products from farms to markets.

5. **Demand forecasting and planning:** Technology solutions enable more accurate demand forecasting and planning in agribusiness. Advanced analytics and machine learning algorithms can analyze historical data, market trends, and external factors to predict demand patterns. This helps farmers and agribusinesses adjust production levels, manage inventory, and coordinate supply chain activities to meet market demands effectively.

6. **Collaboration and communication:** Technology tools improve collaboration and communication among various stakeholders in the supply chain. Online platforms, mobile applications, and cloud-based systems enable seamless sharing of information, such as product specifications, quality standards, and pricing. This facilitates better coordination, reduces communication gaps, and fosters stronger relationships between farmers, suppliers, distributors, and retailers.

7. **Quality control and certification:** Technology adoption allows for more effective quality control and certification processes in agribusiness. Automated inspection systems, image recognition, and machine vision technologies can assess the quality attributes of agricultural products, ensuring compliance with industry standards. This reduces the reliance on manual inspection, improves accuracy, and expedites the certification process.

By leveraging technology in supply chain management, agribusinesses can optimize operations, reduce costs, minimize wastage, ensure food safety, and enhance customer satisfaction. It leads to a more efficient and resilient agricultural supply chain, benefiting both producers and consumers.

Barriers in the implementation of Technology adoption in Agribusiness

Also, despite recent advances in different technologies that support Agribusiness, companies must take measures to mitigate several underlying risks:

» **Cyberattacks:** As supply chains are becoming more digitalized and rely on real-time communication and coordination of many different devices (sensors, robots, drones, etc.), there is a heightened concern about cybersecurity. For instance, because Industry 4.0 involves many devices communicating via different operating systems and different information systems, digitalized supply chains can be vulnerable to cyberattacks. These attacks can include industrial espionage, IP leakage, or even production sabotage. In some cases, cyberattacks can bring an entire logistics/transportation system to its knees.

» **Faulty data:** Smart devices are capable of sensing, collecting, sharing, analyzing data, etc. so that they can facilitate many supply chain operations smoothly. However, if the system is hacked or if the system is not functioning properly, disasters ensue as we witnessed two crashes of Boeing's 737 Max in 2019 due to faulty readings of a key sensor. It is simply too risky to rely on smart devices to manage a supply chain; hence, companies must develop a fool-proof protocol with human interventions.

» **Safety regulations:** Many firms are exploring the use of advanced robotics and automated guided vehicles in manufacturing, robotic systems in automated storage and retrieval warehouse operations, and autonomous trucks and drones in delivery operations. Some standard safety guidelines and regulations must be developed to ensure worker and public safety. For instance, the operations of Heathrow and Gatwick airports in London airports were disrupted after unauthorized drones were flying within restricted airspace in 2018.

» **Privacy issues:** Many sensors record visuals and sounds of different movements of different people in many places. For example, many smart shelves record shoppers' demographic data and their shopping habits within a shop. As such, there is a legitimate concern over privacy issues about this massive data collection. The government must develop regulatory measures to ensure companies collect and store this data properly for public safety.

Conditions needed to enable smart agriculture businesses:

For Smart Agribusiness to have a positive effect on the agriculture sector throughout a region, a few fundamental prerequisites or restrictions must be overcome. They include:

1. Rural infrastructure and connectivity;
2. Affordable Internet access;
3. widespread digital literacy;
4. A culture of rural entrepreneurship; and
5. Assistance from the government and institutions.

Another crucial requirement for Smart Agribusinesses, particularly in ICs and rural regions, is digital literacy. Not only is basic reading necessary for the development of Smart Agriculture but also data processing and communication abilities. Digital literacy can encompass the most basic IT abilities (such as utilizing a mobile phone, a weather station, or buying or selling items online) all the way up to the most complex ones (e.g. fertilization planning using satellite imagery). Digital education needs to advance quickly in communities where even the most fundamental abilities are inadequate (as in the case of ICs). When reflecting on the global advancement of digital literacy, it is important to keep in mind that, due to gender-specific disparities in digital literacy and technology use, female and male workers are likely to be affected differently. The development of digital skills is also being held back, especially in rural areas. A model of digital skills training for farmers and other vulnerable groups along the agribusiness value chains is required in order to teach them how to evaluate and use best practices and technologies for their farm company.

Opportunities Related to adoption of Technology in Agribusiness Activities:

The problems that occur at various stages of the value chains for agriculture give rise to opportunities. The four areas of agriculture—are crops, livestock, fisheries, and non-food products. Each offers several potentials to advance the current state-of-the-art. Smart Agribusiness offers these opportunities. Agriculture 4.0 is a market trend that can't be stopped and can't be turned around. Using personal cell phones, these technologies might connect people's current connectivity. Smartphones, IoT, and AI have helped to facilitate the gradual integration of communication and information technologies into the psycho-social structures of people, cities, and industries. Because of how simple and affordable these technologies are to use, there is a tremendous amount of untapped potential. Because of the current global population growth, resource scarcity, and climate change, agriculture is currently severely stressed. Agriculture 4.0 can be helpful in this case since it boosts crop output, enhances crop management choices, lessens the environmental effect of agricultural practices by using fewer chemicals, and lowers expenses for things like power, water, and gasoline. In order to ensure farmers' livelihoods and food security in a changing climate, climate-smart agriculture aims to adapt to and mitigate the effects of climate change by sequestering carbon in the soil, reducing greenhouse gas emissions, and raising the productivity and profitability of farming systems. IoT and precision farming strategies will converge to create a network of networked new farming equipment.

Conclusion

For technology adoption in agribusiness to be successful, a rural entrepreneurial culture is particularly crucial. The formation, development, and scaling-up of "digital start-ups," especially in the agriculture and food industry, is being supported by an expanding number of global initiatives. A long-term political and practical approach, beginning with suitable teaching in schools, is required to create a sustainable entrepreneurial culture for Smart Agribusiness. It requires an enabling environment that promotes risk-taking, trust-based relationships between stakeholders, financial opportunities, professional services, a sustainable digital ecosystem, the availability of necessary skills, and an attitude of sharing or 'open innovation'. If rural women and youth actively participate in the digital ecosystem, smart agribusiness programs and investments will be more long-lasting.

This can be accomplished, for instance, by enhancing their skills and encouraging entrepreneurship, guaranteeing fair access to services and resources that are productive, enhancing involvement in decision-making, and better addressing socio-cultural concerns in the policy.

References

Chavas, J. P., & Nauges, C. (2020). Uncertainty, learning, and technology adoption in agriculture. *Applied Economic Perspectives and Policy*, *42*(1), 42-53.

Ding, Y., Jin, M., Li, S., & Feng, D. (2021). Smart logistics based on the internet of things technology: an overview. *International Journal of Logistics Research and Applications*, *24*(4), 323–345. https://doi.org/10.1080/13675567.2020.1757053

Espíndola, O., Chowdhury, S., Beltagui, A., & Albores, P. (2020). The potential of emergent disruptive technologies for humanitarian supply chains: the integration of blockchain, Artificial Intelligence and 3D printing. *International Journal of Production Research*, *58*(15), 4610–4630. https://doi.org/10.1080/00207543.2020.1761565

Fisher, D. K., Norvell, J., Sonka, S., & Nelson, M. J. (2000). Understanding technology adoption through system dynamics modeling: implications for agribusiness management. *The International Food and Agribusiness Management Review*, *3*(3), 281-296.

Humayun, M., Jhanjhi, N., Hamid, B., & Ahmed, G. (2020). Emerging Smart Logistics and Transportation Using IoT and Blockchain. *IEEE Internet of*

Things Magazine, *3*(2), 58–62. https://doi.org/10.1109/iotm.0001.1900097

Kamble, S. S., Gunasekaran, A., & Sharma, R. (2020). Modeling the blockchain enabled traceability in agriculture supply chain. *International Journal of Information Management*, *52*, 101967.

Kumar, A., Srivastava, S. K., & Singh, S. (2022). How blockchain technology can be a sustainable infrastructure for the agrifood supply chain in developing countries. *Journal of Global Operations and Strategic Sourcing*.

Kumar, S., Raut, R. D., Nayal, K., Kraus, S., Yadav, V. S., & Narkhede, B. E. (2021). To identify industry 4.0 and circular economy adoption barriers in the agriculture supply chain by using ISM-ANP. *Journal of Cleaner Production*, *293*, 126023.

Saurabh, S., & Dey, K. (2021). Blockchain technology adoption, architecture, and sustainable agri-food supply chains. *Journal of Cleaner Production*, *284*, 124731.

Yamoah, F. A., & Kaba, J. S. (2022). Integrating climate-smart agri-innovative technology adoption and agribusiness management skills to improve the livelihoods of smallholder female cocoa farmers in Ghana. *Climate and Development*, 1-7.

Yengoh, G. T., Armah, F. A., & Svensson, M. G. (2010). Technology adoption in small-scale agriculture.

Web References

https://agribusinessedu.com/what-is-smart-agribusiness/

https://www.farmmanagement.pro/modern-agricultural-technology-adoption-and-its-importance/

https://www.bcg.com/publications/2022/agribusiness-and-the-shift-to-sustainable-farming

10 Consumer Trends and Market Analysis in Agribusiness

Dr. Ruqaiya Bano

Assistant Professor, Integral Institute of Agricultural Sciences and Technology (IIAST), Integral University Lucknow

Email: ruqaiyabanoaem@gmail.com

Abstract

Agriculture has transformed into agri-business and grown into a big and intricate industry that now encompasses everyone involved in getting food and fibre to consumers. Agribusiness encompasses not just individuals who cultivate the land but also the people and businesses who supply the inputs, process the output, produce the food products, transport the food products, and sell the food products to customers.

Many consumers are shifting towards organic and locally sourced produce and meat, as well as sustainable agriculture practices. In response to these consumer trends, companies in the agribusiness industry are adapting their business models and production methods. At the same time, companies are also investing in transparency and traceability. Consumers want to know where their food comes from, how it was produced, and how it impacts the environment. As a result, many companies are now providing detailed information on their products, including the source of the ingredients and the production methods used.

Finally, there is increased competition in the agribusiness industry. With the rise of organic and locally sourced food, as well as sustainable agriculture practices, many small businesses and startups are entering the market. This competition is driving innovation and allowing consumers to have more choices in the products they buy. Companies are adapting their business models and production methods to meet the demands of this new market. As a result, consumers have more choices when it comes to their food, and the agribusiness industry is becoming more sustainable and transparent. This paper aims to analyse the changing consumer trends and their impact on agribusiness sector.

Keywords: Market Analysis, Consumer Trends, Sustainability, Organic and Local Food, Agribusiness

Introduction

Agribusiness is a sector that deals with agriculture and farming, which plays a significant role in the global economy. Agribusiness is "The sum total of all operations involved in the manufacture and distribution of farm supplies; production operations on the farm; and the storage, processing, and distribution of the resulting farm commodities and items" (Davis &Goldberg, 1956).

The food industry is a crucial element of the agribusiness sector, as it involves the production, processing, and distribution of food products. Over the years, there has been a significant transformation in the agribusiness sector, and the consumer trends have shifted dramatically.

Agriculture has transformed into agri-business and grown into a big and intricate industry that now encompasses everyone involved in getting food and fibre to consumers. Agribusiness encompasses not just individuals who cultivate the land but also the people and businesses who supply the inputs, process the output, produce the food products, transport the food products, and sell the food products to customers. These changes have led to new opportunities for businesses, as well as challenges.

Transformation of Agriculture into Agribusiness

Agriculture was a self-sufficient business at the beginning of the 19th century. The average farming family made their own food, fuel, home, draught animals, feed, tools, utensils, and even clothing. Only a small number of necessities needed to be bought off the farm or bartered for. Practically all tasks related to the production, processing, storage, and distribution of farm products were handled by the farm family (Desai,1974; Hamilton, 2016). In recent years, this has seen significant modification. Today's agriculture is quickly evolving into a technology- and market-driven sector that encompasses both advanced agriscience and agribusiness in addition to agricultural output.

The agribusiness approach is a way of looking at farming issues in a fresh and all-encompassing way. This strategy has had the advantage of freeing up farm labour for use in new non-agricultural occupations, such as the military forces during conflicts. As a result, there has been significant economic growth, advancements, and a rise in living standards.

The several million farm units and the few thousand business units that make up the agribusiness are each separate, self-sufficient entities. Agribusiness is the culmination of hundreds of committees, conferences, quasi-research groups, commodity organisations, trade associations, farm organisations, and commodity organisations, each of which focuses on a different set of interests (Desai,1974; Van Fleet, 2016).

There are many advantages to the transition from agriculture to agribusiness. These include lessening the amount of labourers' work, improving nutrition, increasing peoples' mobility, and providing healthier, more fibrous meals. For the past 150 years, the country's economic growth and progress have been based on the release of farm labour and the introduction of new, off-farm jobs. Increased worker productivity has been essential to this growth and progress, which in turn encourages creativity, new product development, and riches. This translates into increased consumer spending power, new factories, risk capital, and jobs.

Consumer Trends in Agribusiness

Consumer behaviour is a crucial aspect of market analysis and often determines the success of businesses. In the agribusiness sector, the consumer trends that have emerged over the years include the following:

1. Healthy Eating Habits

Consumers' purchasing and eating patterns are changing globally as a result of their evolving interests and lifestyles. Consumers are becoming increasingly health-conscious, and they are more aware of the health benefits of the foods they consume. There is a high demand for fresh, organic, and locally grown produce, free from pesticides, hormones, and other artificial additives.

Consumer food consumption and purchasing patterns are evolving globally as a result of people's evolving preferences and lifestyles. While not all trends are intended to be covered in detail, several of the most significant ones from the perspective of food safety are examined in this brief. increasing interest in plant-based foods is being driven by concerns about healthier diet options and environmental sustainability. This industry is quickly increasing to include plant-based substitutes for meat, dairy, eggs, and seafood. Edible insects and other alternative food sources like seaweeds or macroalgae are also gaining popularity. Urban agriculture has developed and grown as a result of the rapid urbanisation and the necessity for regional, sustainable food production.

The vigorously promoting nutrition education helps people be more objective in assessing their own eating habits and in examining things they believe to be healthy. The services of nutritionists who can suggest a suitable diet are available to customers with unique nutritional needs (Voinea et. al, 2019)

Numerous studies have shown that consumers are very interested in food products with eco-friendly qualities and certifications. This is largely due to the fact that these products contain significant amounts of nutrients that are good for health, first strengthening the immune system and lowering the risk of food allergies (Ricci et. al, 2018)

2. Sustainability and Environmental Consciousness

Improving food security and quality in the global food system depends increasingly on sustainable food production and consumption. The structure of international trade flows, as well as the sustainability of the ecosystem and environment, are greatly influenced by consumer food consumption patterns in India, a rapidly developing country with the largest population and one of the largest consumer markets in the world.

Consumers are more conscious of the environmental impact of food production and are inclined towards sustainable farming and animal welfare. They are seeking out eco-friendly products and packaging, and there is a growing demand for environmentally friendly farming practices and product certification (Zheng et. al, 2018).

The establishment of a sustainable food system may be aided by the adoption of healthy eating habits based on minimally processed natural foods. The choice of food packaging has been significantly influenced by attitudes towards the environment. More optimistic perspectives promoted the reduction of packaging waste. Environmental initiatives that concentrate on changing mindsets and using food packaging that is environmentally friendly are so necessary (Jeżewska-Zychowicz, 2015).

Finally, there is a growing concern about the impact of food production on the environment. Consumers want to know that their food is produced in a way that is sustainable and does not harm the environment. This trend has led to an increased focus on sustainable agriculture practices in the agribusiness industry.

3. Organic and Local Food

Organic and locally sourced food has become increasingly popular in recent years. Over the past few decades, consumers' interest in regional and organic food production has grown. Due to increasingly complicated, globalised food chains and news reports about food scandals around the world, there is a greater desire for greater openness in the food manufacturing process (Adams & Salois, 2010).

Over the past few decades, organic food has gained appeal and has become widely available in most large supermarket stores. Food produced locally, however, is another trend that is gaining popularity with consumers. The distinction between locally grown and organically grown goods can often be very unclear. Organic food is produced without the use of artificial pesticides, herbicides, fertilisers, genetically modified organisms, chemical preservatives, radiation, or antibiotics in animal feed. In addition to maintaining the land, organic farming must also treat cattle humanely. The definition of local food is less rigid. It may or may not be organic, but it is always bought close to the factory (Hempel,2016).

People are concerned about the impact of pesticides and fertilizers on the environment and the health of the people consuming the food. As such, many consumers are willing to pay a premium for organic, locally sourced produce and meat.

4. Meat Alternatives

In recent years, the number of people seeking meat alternatives has grown significantly due to concerns about health, animal welfare, and environmental sustainability issues. The popularity of alternative meat products has led to an increase in investment and innovation in the industry to meet the growing demand. Being a vegetarian or vegan is no longer a stigmatized, niche lifestyle but a mainstream lifestyle. One of the primary reasons for people to switch to meat alternatives is health consciousness. The rise of lifestyle diseases such as obesity, heart disease, and diabetes is prompting people to adopt healthier diets. The shift towards plant-based diets is driven by the perception that meat alternatives are healthier, contain fewer additives, and are less dangerous for human consumption. A study conducted on 11,000 adults in the US found that consumers who buy plant-based meat alternatives are health-conscious, concerned about the environment, and more inclined to express dietary preferences (Dunne et al. 2018).

Another significant reason that drives consumers to opt for meat alternatives is environmental sustainability and ethics. The livestock industry, which is responsible for the production of meat products, accounts for nearly 18% of global greenhouse gas emissions, more than the transport sector (FAO, 2013). The industry requires a substantial amount of land use, water consumption, and contributes to deforestation, soil degradation, and pollution. The public is often alarmed by the negative impact of intensive animal farming on the environment. Consequently, this leads to consumers choosing alternative meat products as they believe that it is a more resource-efficient and sustainable form of protein for the future. One of the most critical factors driving consumers towards meat alternatives is animal welfare. The meat industry has long been associated with inhumane treatment of animals. The exposure to animal welfare concerns, such as caged and cramped farms, slaughter, and transportation conditions of the animals from farmers to processing or packaging sites, makes the general public more.

Market Trends in Agribusiness

The global agribusiness market is undergoing substantial structural changes with technical advancement, implementation of advanced machinery, genetically modified seeds and expanding access to global markets. Growing internet connectivity in rural areas gives farmers access to data on seeds, crops, pest management techniques, weather forecasts, fertilisers, and market pricing, assisting them in making educated decisions that have been demonstrated to be successful.

Shifting preferences of consumer to vegan food varieties, growing environmental awareness among consumers, demand for sustainability and traceability, and preference for organic and healthy and local food during resurging virus conditions are noted as key trends shaping the end-use food and beverages market. However, prolonged Russia-Ukraine war conditions further worsen COVID-induced supply disruptions, Climate change leading to unpredictable weather, and burgeoning labour and processing costs are hindering market growth.

The agricultural sector is going through a period of considerable change on many different levels. The agribusiness sector is in the midst of profound shifts and dynamic trends that will continue to influence how food is grown, raised, and produced for years to come, including shifting consumer food preferences and related shifts towards "clean label" products, sustained low commodity prices for farmers, and the recent surge in AgTech development and adoption.

Agribusiness Market Overview

In India, agriculture provides the main means of subsistence for around 58 percent of the people. Gross value added (GVA) for agriculture, forestry, and fishery was calculated at Rs. 19.48 lakh crore (US$ 276.37 billion) in FY2020 (PE). GVA growth in the agricultural and related sectors was 4% in FY20.

The Indian food industry is poised for huge expansion, increasing its contribution to the global food trade every year because of its enormous value-adding potential, notably within the food processing sector.

The food and grocery sector in India is the sixth largest in the world, with 70% of sales coming from retail. One of India's largest sectors, the food processing sector ranks fifth in terms of production, consumption, export, and predicted growth. It accounts for 32% of the nation's overall food market.

In the Indian agribusiness sector, the number of MSMEs is growing exponentially; from a total count of 9,000 (including Registered and Unregistered MSMEs), it has increased to a total of 55,400 MSMEs, significantly boosting the segment's growth in terms of revenue and social and environmental impact.

Market Opportunities in Agribusiness Sector

1. Production Sector

In the upcoming years, India's overall food output is anticipated to grow at a very good rate. Consequently, there is a chance for significant expenditures in technologies, personnel, and tools related to food and food processing.

Canning, dairy and food processing, speciality processing, packaging, frozen food/refrigeration, and thermo processing are a few examples of the key activities in this field. Fisheries, Milk and Milk Products, Meat and Poultry, Packaged/Convenience Foods, Alcoholic Beverages, Soft Drinks, and Grains are among the food categories that are included.

Another rapidly expanding sector of this economy is health foods and health food supplements. The government of India has given the food processing industry sector great attention and has implemented a variety of tax breaks and incentives to promote commercialization and value addition. To upgrade its infrastructure for agricultural products, India needs sophisticated facilities for germplasm and seed multiplication, agricultural inputs (seeds, plant nutrients, and plant protection chemicals), irrigation, controlled production facilities, farm mechanisation, etc.

2. Processing Sector

Due to the crucial connections and synergies that the food processing industry fosters between the two pillars of our economy, industry and agriculture, it is of utmost importance for India's development.

To attain favourable terms of trade for Indian agriculture in both domestic and international markets, rapid growth in the food processing sector and concurrent improvement in the development of the value chain are equally important.

3. Distribution Sector

India is the second-largest food production in the world, but it lacks the infrastructure and storage facilities needed to transport food in a chilled state. The areas of procurement and delivery systems, pre-cooling facilities, refrigerated trucks, cold stores/controlled ambience stores, warehouses, and traceability difficulties, among others, offer tremendous investment possibilities.

For significant backward and forward integration, an integrated supply chain including warehousing, cold chain, and transportation from the farm gate to the food plate is required. To maximise the effectiveness of utilisation, supply networks must be designed such that they may be used year-round for a variety of goods.

Conclusion

In response to these consumer trends, companies in the agribusiness industry are adapting their business models and production methods. Many are shifting towards organic and locally sourced produce and meat, as well as sustainable agriculture practices. At the same time, companies are also investing in transparency and traceability. Consumers want to know where their food comes from, how it was produced, and how it impacts the environment. As a result, many companies are now providing detailed information on their products, including the source of the ingredients and the production methods used.

Finally, there is increased competition in the agribusiness industry. With the rise of organic and locally sourced food, as well as sustainable agriculture practices, many small businesses and startups are entering the market. This competition is driving innovation and allowing consumers to have more choices in the products they buy. The consumer trends and market analysis in agribusiness are changing rapidly, driven by a growing concern for the environment, animal welfare, and sustainable agriculture practices. Companies are adapting their business models and production methods to meet the demands of this new market. As a result, consumers have more choices when it comes to their food, and the agribusiness industry is becoming more sustainable and transparent.

In conclusion, the agribusiness sector is a dynamic and challenging industry that requires businesses to stay up-to-date with the latest consumer trends and market analysis. The consumer trends in agribusiness include healthy eating habits, sustainability and environmental consciousness, and digitalization and convenience. In order for businesses to succeed in the agribusiness sector, they must adapt to changing consumer trends and market dynamics, and be innovative in their production and distribution network.

References

Voinea, L., Vrânceanu, D. M., Filip, A., Popescu, D. V., Negrea, T. M., & Dina, R. (2019). Research on food behavior in romania from the perspective of supporting healthy eating habits. *Sustainability*, *11*(19), 5255.

Ricci, E. C., Banterle, A., & Stranieri, S. (2018). Trust to go green: an exploration of consumer intentions for eco-friendly convenience food. *Ecological economics*, *148*, 54-65.

Zheng, Q., Wang, H. H., & Lu, Y. (2018). Consumer purchase intentions for sustainable wild salmon in the Chinese market and implications for agribusiness decisions. *Sustainability*, *10*(5), 1377.

Jeżewska-Zychowicz, M., & Jeznach, M. (2015). Consumers' behaviours related to packaging and their attitudes towards environment. *Journal of Agribusiness and Rural Development*, *37*(3).

Hempel, C., & Hamm, U. (2016). How important is local food to organic-minded consumers?. *Appetite*, *96*, 309-318.

Adams, D. C., & Salois, M. J. (2010). Local versus organic: A turn in consumer preferences and willingness-to-pay. *Renewable agriculture and food systems*, *25*(4), 331-341.

Dunne, L. E., Fleck, E. A., Daley, C., Gwin, S. E. (2018). Safe and sustainable: User motivations for buying plant-based meat alternatives. Environmental Management, 62, 128-137.

FAO. (2013). Tackling climate change through livestock - A global assessment of emissions and mitigation opportunities. Rome, Italy: Food and Agriculture Organization of the United Nations.

Markets and Markets. (2021). Plant-based Meat Market by Source (Soy, Wheat, Pea, and Mycoprotein), Product (Burger Patties, Sausages, Strips & Nuggets, Meatballs), Type (Beef, Chicken, Pork, Fish), Process (Processed and Uncut), and Distribution Channel—Global Forecast to 2026. Retrieved from https://www.marketsandmarkets.com/Market-Reports/plant-based-meat-market-44922705.html

Yogesh, M. S., & Chandrashekar, H. M. (2014). Opportunities for strengthening of agribusiness sectors in rural India. *International Journal of Management Research and Reviews*, *4*(4), 498-508.

Hamilton, S. (2016). Revisiting the history of agribusiness. *Business history review*, *90*(3), 541-545.

Desai, D. K. (1974). Evolution of a concept of agribusiness and its application. *Indian Journal Of Agricultural Economics*, *29*(4), 32.

Davis, J. H., & Goldberg, R. A. (1957). Concept of agribusiness.

Van Fleet, D. (2016). What is agribusiness? A visual description. *Amity Journal of Agribusiness*, *1*(1), 1-6.

11

Farmer Cooperatives and Collective Marketing

Richa Khushwaha

Ph.D. Scholar, Department of Business Management, Dr Yashwant Singh Parmar University of Horticulture and Forestry, Nauni.

Abstract

The productivity of small and marginal farmers is crucial to India's future food security and sustainable agricultural growth. Small farmers have a number of challenges, such as a lack of access to high-quality inputs, finance options, public resources, technology, limited output volume, an unreliable market, and unstable revenue. Agricultural co-operatives are farmer-owned cooperatives whose main goal is to boost member farmers' output and earnings by improving their connections to the financial, agricultural input, information, and output markets. Various categories of co-operatives exist in India mainly production cooperatives, marketing cooperatives, service cooperatives and allied service cooperatives. Marketing cooperatives include organisations like agricultural marketing organisations and consumer cooperatives that market agricultural products. By generating income for members, reducing poverty, empowering women, fostering social integration, establishing rural markets, developing human capital, and assisting members in organising for sustainable livelihoods, cooperatives play a key role in the overall economic and social development of society.

Keywords: Cooperative movement, NCDC, Structure, Democratic Governance, NAFED

Introduction

According to International Co-operative Alliance- An autonomous association of persons united voluntarily to meet their common economic, social and cultural needs and aspirations through a jointly owned and democratically controlled enterprise.

History of co-operatives in India

The cooperative movement in India was formally founded in 1904. It was extremely significant in growth of the agricultural industry, particularly in supporting the tiny, marginal, and poorer segments of society. The government immediately after independence recognized the cooperative sector as the third economic sector to complement the public and private sectors, and in 1950, cooperatives were included in the National Development Plans. The First Plan (1951–1956) advocated for the formation of cooperative marketing groups and the training of individuals participating in cooperative growth. The All-India Rural Credit Survey Committee emphasised the necessity for expanding cooperative credit's role for diversification into different economic operations and proposed an integrated approach to cooperative credit in 1954. With a focus on storage and warehousing services, the Second Plan aimed to increase the cooperatives' activities. A National Policy on Cooperatives was created by the National Development Council in 1958. In the Third Plan, it was suggested that the cooperatives engage in spinning and weaving as well as other agro-based businesses including the processing of agricultural products like sugarcane, cotton, milk, etc. To develop the skills of the personnel, office holders, and farmer members, a cooperative training college was formed in Pune with numerous regional centres. The Fourth Plan placed a strong emphasis on the fusion of cooperative systems. Restructuring marketing and consumer organisations to assist agricultural development was a key component of the Fifth Plan. The Sixth Plan sought to expand activities into food processing, fishing, poultry, dairy farming, and other fields while transforming village primary societies into multifunctional societies. With a focus on creating jobs and reducing poverty while building the consumer cooperative movement, particularly in the North-Eastern Region, the Seventh Plan placed a strong emphasis on the involvement of the general public in achieving the core goals. The Eighth Five Year Plan (1992–1997) placed a strong emphasis on boosting economic activity and establishing job possibilities for institutional setups that could govern themselves, control themselves, and rely on themselves. Since the Ninth Plan (1997-2002), cooperatives have not been specifically mentioned as a part of the Plan. Due to the fact that cooperation is a state-level issue, the Multi-State Cooperative Societies Act of 2002 encouraged the coexistence of prior laws with parallel cooperative legislation for self-sufficient cooperatives. According to the National Cooperative Policy of 2002, cooperatives should be encouraged to grow in all areas, be responsible to its members, and significantly impact the national economy. The NCDC Amendment Act of 2002 has made it possible for it to cover industrial, livestock, and notified services, as well as directly finance cooperatives in exchange for adequate security. (Sahoo et al. 2020)

National Cooperative Development Corporation: -

In 1963, a parliamentary act created the National Cooperative Development Corporation (NCDC) as a statutory corporation under the Ministry of Cooperation.

Functions of NCDC-

Planning, promoting, and funding programmes for the production of agricultural products, food products, certain other notified commodities, such as fertilisers, insecticides, agricultural machinery, lac, soap, kerosene oil, textile, rubber, etc., supply of consumer goods, and collection, processing, marketing, storage, and export of minor forest produce through cooperatives, in addition to income-generating streams of activities like poultry farming. NCDC Act has also been revised, which would broaden the Corporation's operational region, support more varieties of cooperatives, and strengthen its financial foundation. NCDC will now be able to provide funding for projects in the rural industrial cooperative sectors and for specific services in rural areas that have been approved, such as water conservation, irrigation, micro irrigation, agri-insurance, agro-credit, rural sanitation, animal health, etc. Advances in loans and grants are made to state governments for the purpose of funding primary and secondary cooperative societies as well as directly to the national level and other societies with objectives that go beyond a single state. As long as all requirements are met, the Corporation may now directly support projects under one of its many aid programmes.

The significance of farmer cooperatives in the agricultural sector stems from their ability to empower farmers, enhance their market power, and promote their economic and social well-being. Here are some key aspects that highlight the significance of farmer cooperatives:

1. ***Collective Bargaining Power***: Individual farmers, who frequently work on small or medium-sized farms, can pool their resources in farmer cooperatives in order to collectively negotiate better prices and terms for their produce. Cooperatives can attain economies of scale and compete more successfully with larger agricultural competitors by pooling their production quantities and market access. (Rananavare, 1964; Ortaman and King, 2007).

2. ***Market Access and Market Development***: By offering group marketing services,

farmer cooperatives help their members gain access to the market. They support farmers in overcoming obstacles to reaching consumers, gaining access to distribution networks, and fulfilling market demands. Cooperatives can establish connections with buyers, look into untapped market potential, and manage logistics and distribution on their members' behalf. (Kumar *et al.*, 2015; Nwankwo *et al.*, 2016)

3. ***Risk Management***: Cooperative marketing enables farmers to control and lessen the risks brought on by volatile markets, shifting prices, and erratic demand. Cooperatives can offer stability and a safety net for their members by distributing the risk across a wider population. They can also provide services and tools for risk management, like crop insurance, storage space, and forward contracts. (Dubey *et al.*, 2009)

4. ***Access to Inputs and Services***: Collective purchases of agricultural inputs like seeds, fertiliser, machinery, and equipment are frequently made possible by farmer cooperatives. Cooperatives can source high-quality inputs, negotiate favourable pricing, and save their members money by using the combined purchasing power of its members. They might also provide services to improve members' knowledge and abilities, like technical support, training, and information sharing. (Lawal *et al.*, 2017)

5. ***Knowledge Sharing and Capacity Building***: Farmer cooperatives provide members with a forum for information exchange, skill improvement, and capacity growth. They give farmers the chance to share knowledge, best practises, and innovations. Farmers can improve their knowledge of agricultural practises, market trends, and business management through training programmes, workshops, and cooperative meetings. (Ibitoye and Stephen, 2012; Olaore and Alao, 2019)

6. ***Social and Economic Development***: Farmer cooperatives support the social and economic growth of rural areas. through enhancing farmers' economic viability and giving them more authority. They establish a sense of ownership and shared prosperity through developing community togetherness, cooperation, and group decision-making. (; Mhembwe and Dube, 2017)

Structure and Functioning of Farmer Cooperatives

The organizational structure and governance models of farmer cooperatives are designed to ensure the active participation and empowerment of farmer-members. Here is a detailed examination of these aspects:

1. ***Membership***: Farmers themselves, who make up the majority of members, own and run farmer cooperatives. Individuals or organisations involved in agricultural production are eligible to join, and in some situations, this may also include other stakeholders like rural communities or agricultural employees. A member's participation in and contribution to the cooperative entitles them to perks and dividends as well as voting rights.

2. ***Democratic Governance***: Farmer cooperatives follow democratic principles, where decisions are made collectively and members have equal voting rights. General assemblies or meetings are held regularly, where members discuss important matters, elect board members, and vote on key issues affecting the cooperative. This democratic governance structure ensures that the decisions and policies of the cooperative align with the interests and needs of the farmer-members.

3. ***Board of Directors***: The members elect the board of directors, which represents their interests in the management of the cooperative. The cooperatives by laws specify the number of directors and their tenure of office. It makes judgements about things like pricing, marketing plans, financial choices, and risk management. In order to ensure accountability and transparency, the board serves as a liaison between the members and the management.

4. ***Executive Management***: Farmer cooperatives frequently have a skilled management group in charge of the cooperative's daily operations. The board of directors appoints the general manager or CEO who leads the executive management team.

5. ***Committees and Subcommittees***: In order to concentrate on particular aspects of the cooperative's operations, farmer cooperatives may create committees or subcommittees. These committees are created based on the members' experience and interests. There are committees for marketing, finances, quality assurance, or sustainability, for instance.

6. ***Financial Structure***: In order to concentrate on particular aspects of the cooperative's operations, farmer cooperatives may form committees or subcommittees. These committees are created depending on the knowledge and interests of the participants. Committees for marketing, economics, quality assurance, or sustainability, for instance, may exist.

7. ***Education and Training***: Programmes that enhance members' knowledge and skills are highly valued by farmer cooperatives. These programmes provide information on financial management, market trends, good governance, and agricultural practises. (Das, 2017)

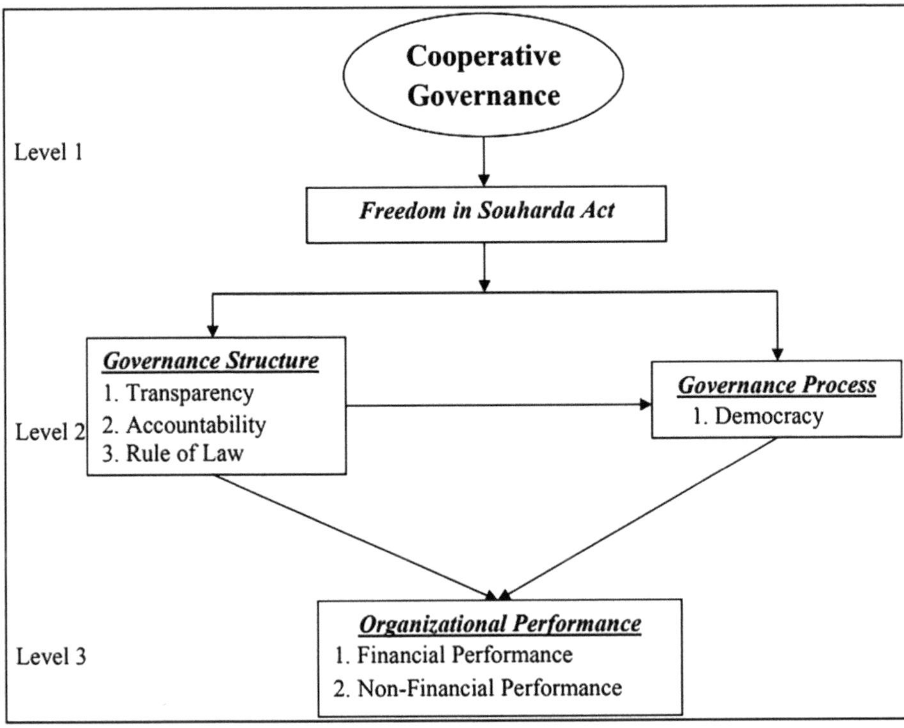

Source- Venkatesha HP. Evaluation of Structure and Process of Governance in Souharda Cooperative: A Conceptual Analysis with Reference to Karnataka, India

Advantages and Benefits of Farmer Cooperatives

1. ***Market Access and Bargaining Power***: Farmers can pool their resources, goods, and services through farmer cooperatives for effective marketing and distribution, giving them access to a larger market. By working together, farmers can receive better deals, access more markets, and get around obstacles that would be challenging to achieve alone. Cooperatives give farmers more negotiating leverage when dealing with consumers, processors, and retailers, improving market conditions and boosting earnings.

2. ***Economies of Scale and Cost Savings***: Farmers can gain from economies of scale by coming together in a cooperative. Individual farmers might not be able to afford cooperative investments in shared infrastructure like processing plants, storage warehouses, and transportation systems. This provides cost savings through group purchasing, bulk sales, and shared expenses, lowering input costs and increasing total profitability for individual farmers.

3. ***Risk Mitigation***: Farmer cooperatives help mitigate risks by pooling resources and diversifying product offerings. By spreading risks across a larger membership base, cooperatives provide individual farmers with a safety net against market fluctuations, weather-related risks, and unforeseen events. Cooperatives often offer risk management services, such as insurance programs and hedging options, to protect farmers' interests and stabilize their income.

4. ***Access to Services and Resources***: Farmer cooperatives give access to a variety of resources and services that individual farmers might not be able to find on their own. These services could include technical support, educational opportunities, marketing assistance, financial support, and cooperatively owned inputs. Individual farmers can strengthen their farming practises, increase productivity, and get access to collective knowledge by utilising the cooperative's networks and resources.

5. ***Collective Decision-Making and Empowerment***: Individual farmers have a forum through farmer cooperatives to actively participate in decisions that have an impact on their livelihoods. Farmers can influence policies, strategic directions, and resource allocation since they have a voice and voting rights in the cooperative's governance. This kind of group decision-making improves the sense of ownership and control that each farmer has over their agricultural operations.

6.***Policy Advocacy and Representation***: Farmer cooperatives play a vital role in advocating for policies and regulations that benefit the agricultural community.

7.***Knowledge Sharing and Collaboration***: Farmer cooperatives encourage communication and cooperation among farmers. Members get the chance to share best practises, learn from one another's experiences, and exchange information. This environment of cooperation fosters creativity, advances farming methods, and raises the general output and sustainability of the agricultural sector. (Abudu, 2020)

Cooperative Marketing

According to the Reserve Bank of India "a cooperative marketing society is an association of cultivators formed primarily for the purpose of helping the members to market their produce more profitably than possible through the private trade".

National Agricultural Cooperative Marketing Federation (NAFED)

In India, the NAFED is the federal body for the top co-operative marketing societies at the state level. It was founded in 1958, and New Delhi serves as its administrative centre. Its main job is to give advice and technical direction to state federations while coordinating their activities. Additionally, the federation engages in export and interstate trade.

Role of farmer cooperatives in collective marketing

Farmer cooperatives play a crucial role in collective marketing and market access for their members. Here's an explanation of their role in these areas:

1. ***Collective Marketing:*** - In order to increase their market power and competitiveness, farmer cooperatives give individual farmers the ability to market their goods together. To facilitate collective marketing, farmer cooperatives do the following:

   a. Pooling Resources: Farmers can aggregate their output by combining their resources, which results in higher output quantities and economies of scale. As a result, cooperatives are able to obtain lower prices, gain access to profitable marketplaces, and draw in larger customer bases.

   b. Efficient Distribution: Cooperatives of farmers manage the transportation and distribution of goods. They speed up the collection, grading, sorting, packaging, and transportation procedures to make sure that goods are delivered to clients quickly and effectively. This effectiveness boosts a product's marketability and lowers transaction costs.

   c. Market Information and Analysis: The members of cooperatives receive market research and analysis. They collect information on customer preferences, price, market trends, and dynamics of supply and demand. Farmers can use this information to decide when to produce, how much to charge, and when to sell their products.

   d. Branding and Quality Control: Farmer cooperatives frequently create powerful brands that stand for the dependability and quality of their goods. Cooperatives increase market demand for their goods by upholding constant quality standards and fostering trust with buyers and customers.

   e. Market Diversification: Cooperatives investigate potential markets and broaden their range of goods. They extend the number of markets available to their members by locating niche markets, value-added goods, and export opportunities. Market diversity helps to lessen reliance on a single market and dangers. (Sehgal, 2014)

2. ***Market Access***: Farmer cooperatives enhance market access for their members,

particularly for small-scale farmers who may face barriers to entry in the market. Here's how farmer cooperatives facilitate market access:

a. Negotiating Power: Farmers who work together as cooperatives have more clout when negotiating with consumers, producers, and merchants. For their members, cooperatives can negotiate better rates, agreements, and market conditions, ensuring a just price for their goods.

b. Compliance with Market Requirements: Cooperatives assist their members in adhering to the demands and standards of the market, such as organic certifications, food safety certificates, and sustainable farming methods. As a result, farmers might gain access to niche markets that have strict requirements for product quality.

c. Market Linkages: Market participants such as wholesalers, retailers, exporters, and processors are connected to and maintained through cooperatives. These connections give their members direct market ties, which lessens the need for middlemen and gives them access to buyers directly.

d. Access to Finance and Credit: Access to loans and financing is frequently made easier for members of farmer cooperatives. In order to assist farmers in overcoming financial obstacles that can restrict their access to markets, they might negotiate better financing conditions, arrange credit facilities, and offer collateral backing.

e. Market Research and Development: For the sake of discovering new market trends, consumer preferences, and business openings, cooperatives invest in market research and development. With the aid of this knowledge, farmers are better equipped to satisfy market demands by modifying their production and marketing tactics. (Kapuka. 2017)

By facilitating collective marketing and improving market access, farmer cooperatives empower farmers to participate in competitive markets, increase their profitability, and enhance their overall economic well-being. These cooperatives act as catalysts for farmers' market success by providing the necessary support, resources, and collective strength to navigate the complex and dynamic agricultural markets.

Challenges and Opportunities in Co-operative Marketing

Cooperative marketing, including branding, product differentiation, and value-added initiatives, presents both challenges and opportunities for farmer cooperatives. Let's examine them in more detail.

Challenges:

1. ***Limited Resources:*** - Resources such as capital, infrastructure, and technical skills are frequently restricted for farmer cooperatives. Financial investments and specialised knowledge are needed to develop efficient branding strategies, implement product differentiation, and launch value-added activities. Cooperatives, especially those with fewer resources, may find it difficult to overcome these restrictions.

2. ***Competing with Established Brands***: For farmer cooperatives, establishing a strong brand presence and competing with well-known companies can be difficult. The market familiarity, consumer loyalty, and marketing budgets of established businesses are frequently higher. It might be challenging for cooperatives to spend in branding and marketing initiatives that will help them stand out from the competition and increase their market share.

3. ***Consistent Quality Control***: For branding, product differentiation, and value-added efforts, maintaining consistent product quality is crucial. To constantly fulfil market criteria, cooperatives need to set up strong quality control methods. It might be difficult to ensure that all members follow quality standards and to execute quality control procedures during the manufacturing and processing stages, especially when working with a diversified membership base.

4. ***Market Access and Distribution***: Farmer cooperatives must build effective distribution channels for their branded and value-added products and secure dependable market access. In addition to tackling logistics and supply chain issues, this calls for building connections with buyers, retailers, and distributors. Construction and maintenance of these market access channels can be challenging, especially when competing with more established and substantial distribution networks.

Opportunities:

1. ***Differentiation and Market Niche***: Farmer cooperatives can differentiate their products and gain access to specific markets through cooperative marketing. Cooperatives can meet consumer expectations for specialised and differentiating products by focusing on distinctive product features like organic, sustainable, or locally sourced. As a result, there are chances to seize niche markets and command high pricing.

2. ***Direct Consumer Relationships***: Farmers and consumers can establish direct relationships because of cooperative marketing. Cooperatives can develop relationships with consumers and convey the worth of their products by setting up retail outlets like farmers' markets, internet platforms, or community-supported agriculture (CSA) programmes.

3. ***Value-Added Initiatives***: To take a bigger piece of the value chain, cooperatives can explore value-added projects like turning raw materials into final products. Because of this, cooperatives may increase their margins, develop new revenue streams for its members, and distinguish their products from the competition. Cooperatives can broaden their market reach and vary their product offers by implementing value-added activities.

4. ***Collective Strength and Collaboration***: Farmer cooperatives can take advantage of their teamwork and strength in numbers to overcome obstacles and seize opportunities. Cooperatives can invest in branding, product development, and marketing activities that would be challenging for individual farmers by pooling resources, knowledge, and skills. Cooperatives can share costs, gain access to bigger markets, and increase their marketing reach through member cooperation.

5. ***Sustainability and Consumer Preferences:*** Farmer cooperatives can benefit from their collective power and ability to work together to overcome challenges and capture opportunities. By combining their resources, expertise, and knowledge, cooperatives may invest in branding, product development, and marketing initiatives that would be difficult for individual farmers to undertake. Through member cooperation, cooperatives can reduce expenses, gain access to larger markets, and broaden their marketing reach.

Role of government agencies, cooperative development organizations, and financial institutions in supporting cooperative growth and success

Government agencies, cooperative development organizations, and financial institutions play vital roles in supporting cooperative growth and success. Here's an overview of their roles:

1. Government Agencies:

» *Policy and Legal Framework*: The development and functioning of cooperatives are supported by the policies and legal framework created by government organisations. They create laws and rules especially for cooperatives, assuring their legal recognition and fostering an atmosphere that is favourable to their development.

» *Financial Support*: Grants, loans, and subsidies from government organisations are frequently used to support cooperatives financially. Cooperatives can use these cash for things like beginning costs, building out infrastructure, working capital, and expansion plans.

» *Technical Assistance*: Governmental organisations provide cooperatives with technical support in areas such cooperative creation, governance structures, business planning, financial management, marketing tactics, and compliance with legal requirements. For the benefit of management and cooperative members, they might offer consulting services, workshops, and training programmes.

» *Market Access*: Governmental organisations facilitate cooperatives' market access by setting up trade shows, exhibits, and buyer-seller gatherings. In addition, they might aid market research, promote cooperative goods both domestically and abroad, and facilitate connections between markets.

» *Policy Advocacy*: Government agencies advocate for policies and regulations that support cooperative development. They represent the interests of cooperatives in policy discussions, engage in dialogue with other stakeholders, and seek to create an enabling environment for cooperative growth.

2. Cooperative Development Organizations:

» Technical Assistance and Training: Cooperative development organizations specialize in providing technical assistance, training, and capacity-building programs to cooperatives. They offer expertise in cooperative governance, business management, financial planning, marketing strategies, value-added initiatives, and sustainability practices. These organizations help cooperatives improve their operational efficiency, enhance their decision-making capabilities, and adopt best practices.

» Networking and Collaboration: Cooperative development organizations facilitate networking and collaboration among cooperatives. They create platforms for cooperatives to share experiences, exchange knowledge, and learn from one another. These networks promote peer learning, facilitate mentorship opportunities, and foster collaboration on joint projects or market initiatives.

» Research and Advocacy: Cooperative development organizations conduct research on cooperative models, trends, and challenges. They generate knowledge that informs policy discussions and advocates for the recognition and support of cooperatives. These organizations contribute to the understanding of cooperative principles and their potential impact on sustainable development.

- Funding and Grants: Some cooperative development organizations offer funding and grants to support cooperative projects, research initiatives, and capacity-building activities. They provide financial resources that can help cooperatives overcome financial barriers and implement innovative projects.

3. Financial Institutions:

- Access to Capital: Financial institutions play a crucial role in providing access to capital for cooperatives. They offer loans, lines of credit, and other financial products tailored to the specific needs of cooperatives. These financial resources help cooperatives with working capital, infrastructure development, equipment purchase, and expansion plans.
- Financial Expertise: Financial institutions possess expertise in cooperative financing and understand the unique financial challenges faced by cooperatives. They can provide advice and guidance on financial management, risk assessment, and investment strategies to help cooperatives make informed financial decisions.
- Risk Mitigation: Financial institutions can assist cooperatives in managing financial risks and uncertainties. They may offer insurance products, risk assessment services, and hedging mechanisms that help cooperatives mitigate potential financial losses.
- Investment Support: Financial institutions can support cooperatives by investing in their projects and initiatives. They may provide venture capital, equity investments, or impact investing opportunities that enable cooperatives to access additional resources for growth and expansion.

Challenges

Farmer cooperatives face various challenges that can impact their operations and growth. Here are some common challenges faced by farmer cooperatives:

1. Internal Conflicts and Governance Issues:

- Lack of Member Engagement: Some cooperatives struggle with low member engagement and participation, leading to a disconnect between the management and the members. This can hinder decision-making, cooperative governance, and the ability to address members' needs effectively.

» Power Struggles: Internal conflicts can arise due to differing interests, disagreements over strategic decisions, or conflicts of personality among cooperative members or leaders. Power struggles can hinder cooperative cohesion, decision-making processes, and overall effectiveness.

» Leadership and Succession Planning: Ensuring capable leadership and effective succession planning is crucial for the long-term sustainability of farmer cooperatives. The lack of skilled leaders or a smooth transition process can disrupt operations and impede the continuity of cooperative activities.

2. Market Dynamics and Competitiveness:

» Market Volatility: Farmer cooperatives often operate in agricultural markets that are subject to price fluctuations, changing consumer demands, and market uncertainties. Adapting to market volatility and effectively managing risks, such as price fluctuations and weather-related challenges, can be a significant challenge for farmer cooperatives.

» Limited Bargaining Power: Farmer cooperatives may face challenges in negotiating favourable terms with buyers, processors, or retailers due to their smaller size and limited bargaining power. This can result in lower prices, unfavourable contracts, or unequal distribution of benefits along the value chain.

» Market Access and Distribution: Gaining access to markets, especially in competitive and concentrated markets, can be challenging for farmer cooperatives. Establishing reliable distribution channels, reaching new customers, and meeting market requirements for quality, packaging, and certifications can pose barriers for cooperatives.

3. Regulatory and Legal Constraints:

» Compliance with Regulations: Farmer cooperatives must comply with various regulations related to cooperative governance, accounting standards, food safety, labelling, and environmental regulations. Ensuring compliance can be complex and time-consuming, requiring resources and expertise.

» Access to Support and Funding: Some farmer cooperatives face challenges in accessing government support programs, cooperative development funds, or financial services. Limited awareness of available resources, complex application procedures, or stringent eligibility criteria can hinder cooperative access to necessary support and funding.

» Evolving Legal and Policy Frameworks: Changes in agricultural policies, trade regulations, and cooperative laws can impact the operating environment for farmer cooperatives. Adapting to new legal requirements or policy shifts can be challenging, requiring cooperatives to stay updated, seek legal advice, and engage in advocacy efforts.

Addressing these challenges requires strategic planning, collaboration, and continuous capacity-building efforts. Farmer cooperatives can overcome these challenges by fostering member engagement, strengthening cooperative governance, enhancing market intelligence and negotiation skills, seeking partnerships and alliances, engaging in advocacy efforts for supportive policies, and investing in training and education to build the capacity of cooperative members and leaders.

Conclusion

Farmer cooperatives play a crucial role in promoting agricultural sustainability and rural development. Here are some key reasons why farmer cooperatives are important in these areas:

1. Sustainable Agricultural Practices:

» Knowledge Sharing and Training: Farmer cooperatives facilitate the exchange of information, best practices, and training among members. This promotes the adoption of sustainable agricultural practices, such as soil conservation, water management, integrated pest management, and agroforestry. Cooperatives can provide technical assistance and support to help farmers implement these practices effectively.

» Scale and Collective Action: Through cooperatives, farmers can pool resources, share costs, and access economies of scale, making it more feasible to invest in sustainable technologies and practices. Cooperatives can jointly invest in infrastructure, equipment, and renewable energy systems, enabling members to adopt sustainable practices collectively.

2. Market Access and Value Addition:

» Collective Marketing: Farmer cooperatives aggregate and market members' produce collectively, enhancing their market access and negotiating power. This allows small-scale farmers to access larger markets, command better prices, and compete more effectively in the agricultural value chain.

» Branding and Product Differentiation: Cooperatives can develop collective brands and labels that signify sustainable and high-quality agricultural products. This differentiation can attract environmentally conscious consumers and premium market segments, generating higher revenues for cooperative members.

» Value-Added Processing: Farmer cooperatives can invest in value-added processing activities, such as food processing, packaging, and storage. This adds value to agricultural products, extends their shelf life, and opens up new market opportunities. Value-added initiatives generate additional income for farmers, supporting their economic sustainability.

3. Rural Development and Community Empowerment:

» Income Generation and Economic Stability: Farmer cooperatives contribute to rural development by generating income and improving the economic stability of farmers. By providing fair prices, reducing marketing costs, and minimizing post-harvest losses, cooperatives enable farmers to earn a decent income and invest in their farms, education, healthcare, and infrastructure.

» Capacity Building and Skills Development: Cooperatives offer training, education, and capacity-building programs to their members, enhancing their agricultural knowledge, business management skills, and entrepreneurial abilities. This empowers farmers to make informed decisions, improve productivity, and diversify their income sources.

» Social Cohesion and Collective Action: Farmer cooperatives foster social cohesion and collective action within rural communities. By bringing farmers together, cooperatives create opportunities for networking, collaboration, and community engagement. They promote democratic decision-making, respect for diversity, and the sharing of benefits among members.

4. Environmental Stewardship and Natural Resource Management:

» Conservation of Biodiversity: Farmer cooperatives can promote biodiversity conservation by encouraging crop diversification, preservation of traditional seed varieties, and the adoption of agroecological practices. This helps preserve genetic diversity, protect ecosystems, and contribute to sustainable food systems.

» Land and Water Management: Cooperatives can support sustainable land and water management practices among their members, such as conservation agriculture, precision irrigation, and watershed management. These practices minimize soil erosion, protect water resources, and contribute to long-term environmental sustainability.

» Climate Change Mitigation and Adaptation: Farmer cooperatives can play a role in climate change mitigation and adaptation. They can promote climate-smart agriculture techniques, such as agroforestry, carbon sequestration, and renewable energy use. Cooperatives can also support members in accessing climate finance and implementing climate resilience strategies.

Overall, farmer cooperatives are instrumental in promoting agricultural sustainability and rural development. They provide a platform for farmers to collaborate, access markets, adopt sustainable practices, and improve their livelihoods. By fostering social, economic, and environmental sustainability, farmer cooperatives contribute to the long-term well-being of rural communities and the agricultural sector as a whole.

Conclusion

Farmer cooperatives play a crucial role in promoting agricultural sustainability and rural development like Sustainable Agricultural Practices (Knowledge Sharing and Training: Farmer cooperatives facilitate the exchange of information, best practices, and training among members. This promotes the adoption of sustainable agricultural practices, such as soil conservation, water management, integrated pest management, and agroforestry. Cooperatives can provide technical assistance and support to help farmers implement these practices effectively), Scale and Collective Action (Through cooperatives, farmers can pool resources, share costs, and access economies of scale, making it more feasible to invest in sustainable technologies and practices. Cooperatives can jointly invest in infrastructure, equipment, and renewable energy systems, enabling members to adopt sustainable practices collectively), Market Access and Value Addition (Collective Marketing: This allows small-scale farmers to access larger markets, command better prices, and compete more effectively in the agricultural value chain ,Branding and Product Differentiation: Cooperatives can develop collective brands and labels that signify sustainable and high-quality agricultural products, Value-Added Processing: Farmer cooperatives can invest in value-added processing activities, such as food processing, packaging, and storage. Value-added initiatives generate additional income for farmers, supporting their economic sustainability), Rural Development and Community Empowerment (Income Generation and Economic Stability: Farmer cooperatives contribute to rural development by generating income and improving the economic stability of farmers. By providing fair prices, reducing marketing costs, and minimizing post-harvest losses, cooperatives enable farmers to earn a decent income and invest in their farms, education, healthcare, and infrastructure, Capacity Building and Skills Development: Cooperatives offer training, education, and capacity-building programs to their members, enhancing their agricultural knowledge, business management skills, and entrepreneurial abilities. This empowers farmers to make informed decisions, improve productivity, and diversify their income sources, Social Cohesion and Collective Action: Farmer cooperatives foster social cohesion and collective action within rural communities. By bringing farmers together, cooperatives create opportunities for networking, collaboration, and community engagement. They promote democratic decision-making, respect for diversity, and the sharing of benefits among members), Environmental Stewardship and Natural Resource Management (Conservation of Biodiversity: Farmer cooperatives can promote biodiversity conservation by encouraging crop diversification, preservation of traditional seed varieties, and the adoption of agroecological practices. This helps preserve genetic diversity, protect ecosystems, and contribute to sustainable food systems, Land and Water Management: Cooperatives can support sustainable land and water management practices among their members, such as conservation agriculture, precision irrigation, and watershed management. These practices minimize soil erosion, protect water resources, and contribute to long-term environmental sustainability, Climate Change Mitigation and Adaptation: Farmer cooperatives can play a role in climate change mitigation and adaptation. They can promote climate-smart agriculture techniques, such as agroforestry, carbon sequestration, and renewable energy use. Cooperatives can also support members in accessing climate finance and implementing climate resilience strategies).

Overall, farmer cooperatives are instrumental in promoting agricultural sustainability and rural development. They provide a platform for farmers to collaborate, access markets, adopt sustainable practices, and improve their livelihoods. By fostering social, economic, and environmental sustainability, farmer cooperatives contribute to the long-term well-being of rural communities and the agricultural sector as a whole.

References

Abudu S and Annate I. 2020. Contributions of cooperative societies to agricultural development in Edo state, Nigeria: A case study of Akoko-Edo local government area. *Agrotechnology* **9**:1-3.

Adefila J and Madaki J. 2014. Roles of farmers' cooperatives in agricultural development in sabuwa local government area of Katsina state, Nigeria. *Journal of Economics and Sustainable Development* **5**:80-87.

Adeyonu A G, Agboola T O, Fadeyi B O, Tijani B A. and Oladapo E F. 2012. Influence of membership of cooperative society on farmers' income in rural areas of Kwara state, Nigeria. *International Journal of Research and Scientific Innovation* **8**: 84-89.

Adugna H. 2013. Co-operative approach to community livelihood improvement: The case of Ada'a district, Oromia regional state, Ethiopia. *International Journal of Development and Sustainability* **2**:2124-45.

Ahmed M H and Mesfin H M. 2017. The impact of agricultural cooperatives membership on the wellbeing of smallholder farmers: empirical evidence from eastern Ethiopia. *Agricultural and Food Economics* **5**:1-20.

Anonymous. 2019. *Agriculture Census 2015-2016*. Ministry of Agriculture & Farmers welfare, New Delhi. 97p.

Anonymous. 2021. Agriculture. National portal of India (Government of India). *www.india.gov.in/topics/agriculture* (11.00AM, 15th January 2022).

Candemir A and Duvaleix S. 2021. Agricultural cooperatives and farm sustainability: A literature review. *Journal of Economic Surveys* **35**: 1118–44.

Chambers R and Conway G. 1992. Sustainable rural livelihoods: Practical concepts for the 21st century. *IDS Discussion Paper* **296**:1-29.

Das S. 2017. Performance of the primary agricultural cooperative societies: Special reference to short term cooperative credit structure in north eastern region. *Amity Journal of Agribusiness* **2**:22-29.

Dev S M. 2012. Small farmers in India: Challenges and opportunities.

http://www.igidr.ac.in/pdf/publication/WP-2012-014.pdf (11.00 AM, 28th January 2022).

Donkor E and Hejkrlik J. 2021. Does commitment to cooperatives affect the economic benefits of smallholder farmers? Evidence from rice cooperatives in the western province of Zambia. *Agrekon* **60**:408-23.

Dubey A K, Singh A K, Singh R K, Singh L, Pathak M and Dubey V K. 2009. Cooperative societies for sustaining rural livelihood: A case study. *Indian Research Journal of Extension Education* **9**:43-46.

FAO. 2002. *Smallholder Farmers in India: Food Security and Agricultural Policy*. FAO Regional Office for Asia and the Pacific, Bangkok. 54p.

FAO. 2011. *Agricultural cooperatives are key of reducing hunger and poverty*. Food and agriculture organization of the United Nations, Rome. *FAO - News Article: Agricultural cooperatives are key to reducing hunger and poverty (google.com)* (4.00PM, 14th January 2022)

Hailemariam M, Yohannes T and Jemaneh M. 2017. Assessment of the role of agricultural cooperatives in achieving socio-economic development in Sululta Woreda, Oromia special zone. *Journal of Biology, Agriculture and Healthcare* **7**:5-14.

Hassang H. 2006. Role of co-operative societies in raising the standard of living of indigenous people of Arunachal Pradesh: A case study of Apatani tribe. Ph D Thesis. Department of Economics, Rajiv Gandhi University, Arunachal Pradesh. 179p.

Ibitoye and Stephen J. 2012. Survey of the performance of agricultural cooperative societies in Kogi state, Nigeria. *European Scientific Journal* **8**:98-114.

Kapuka H H. 2017. The role of co-operatives in economic development in Namibia. MBA Thesis. Department of Political and Administrative Studies, University of Namibia, Windhoek. 110p.

Kareem R O, Arigbabu Y D, Akintaro J A. and Badmus M A. 2012. The impact of co-operative society capital formation: A case study of temidere co-operative and thrift- society, Ijebu- Ode, Ogun State, Nigeria. *Global Journal of Science Frontier Research Agriculture and Veterinary Sciences* **12**:17-29.

Kumar V, Wankhede K C. and Gena H C. 2015. Role of cooperatives in improving livelihood of farmers on sustainable basis. *American Journal of Educational Research* **3**:1258-66.

Lawal K A, Taiwo A O. and Katagum B. 2017. Promoting rural livelihood through cooperative society in Lagos state: A study of farmers multipurpose cooperative societies in Ojo L.G.A of Lagos State Nigeria. *Nigerian Journal of Management Sciences* **6**:254-62.

Mondal E. 2020. The role of cooperatives in agriculture sector.

Mondal E. *2020. The Role of Cooperatives in Agriculture Sector The Role of Cooperatives in Agriculture Sector (indianfolk.com) - Search (bing.com)* (11.00AM, 19th December 2021)

Mhembwe S and Dube E. 2017. The role of cooperatives in sustaining the livelihoods of rural communities: The case of rural cooperatives in Shurugwi District, Zimbabwe. *Jàmbá: Journal of Disaster Risk Studies* **9**:1-9.

Netravathi K. 2018. A study on corporate governance in SSK cooperative society: with special reference to member's satisfaction. *Sumedha Journal of Management* **7**:172-89.

Njagi N M. 2012. Factors influencing performance of co-operatives; A case of Mbeere north sub-county, Kenya. M A Thesis. School of Biological Sciences, University of Nairobi, Kenya.103p.

Nwankwo F, Ogbodo O C. and Ewuim N. 2016. Effect of cooperative type and age on profit performance: A study of cooperative societies in Awka North LGA in Anambra State, Nigeria. *African Research Review* **10**:187-203.

Olaore R A and Alao K O. 2019. Examination of the roles of cooperative society on the promotion of smallscale enterprises in Osun State, Nigeria. *International Journal of Research and Innovation in Social Science.* **3**:364-69.

Oluyombo O O. 2013. Impact of cooperative societies savings scheme in rural finance: some evidence from Nigeria. *Journal of Economics and Business* **11**:77-88.

Ortmann G and King R. 2007. Agricultural cooperatives I: History, theory and problems. *Agricultural Economics Research, Policy and Practice in Southern Africa* **46**:40-68.

Rananavare A. M.1964. The Role of Cooperative Societies in the Economic Development of India. M. Sc. Thesis. Department of Economics, Utah State University, Logan.104p.

Sahoo A K, Meher S K, Panda T C, Sahu S, Begum S. and Barik N.C. 2020. Critical Review on Cooperative Societies in Agricultural Development in India. *Current Journal of Applied Science and Technology* **39**:114-21.

Sehgal G and Mir A. 2014. Cooperative societies-growth and development- A case study of Jammu and Kashmir state. *International Journal of Business Management* 1:2349-3402

Sevinç M R. 2021. Farmers' perception of agricultural cooperatives: the case of Şanlıurfa, Turkey. *Ciência Rural* **51**:1-11

Sifa C B. Role of cooperatives in agricultural development and food security in Africa.

https://www.un.org/esa/socdev/documents/2014/coopsegm/Sifa--Coopsand agric dev. pdf (11.00AM, 20th January 2022)

Tortia E C, Valentinov V. and Iliopoulos C. 2013. Agricultural cooperatives. *Journal of Entrepreneurial and Organisational Diversity* **2**:23-36.

Theresa U E. 2017. Impact of cooperative society in empowerment of rural women: A case of indigenous women of Nimo town, Anambra state Nigeria. *Saudi Journal of Business and Management Studies* **2**:488-99.

Tumenta B F, Amungwa F A. and Nformi M I. 2021. Role of agricultural cooperatives in rural development in the era of liberalization in the North West and South West regions of Cameroon. *Journal of Agricultural Extension and Rural Development* **13**:69-81.

Ubana E A, Salau E S. and Girei A A. 2019. Constraints to performance of the agricultural cooperative societies services to members in the southern agricultural zone of Nasarawa state, Nigeria. *International Journal of Innovative Research and Development* **8**:1-6.

Valentinov V. 2007. Why are cooperatives important in agriculture? An organizational economics perspective. *Journal of Institutional Economics* **3**: 55–69.

Yu L and Nilsson J. 2021. Farmers' assessments of their cooperatives in economic, social, and environmental terms: An investigation in Fujian, China. *Frontier of Environmental Science* **9**:1-11.

Zhang J, Luo J. and Li J. 2021. Agricultural co-operatives participating in supply chain integration in China: A qualitative comparative analysis. *https://journals.plos.org/plosone/article?id=10.1371/journal.pone.0250018* (4.00PM, 12th January 2022).

12

International Trade and Globalization in Agribusiness

Rohit Kumar Sharma[1] and Mohit Kumar[2]

[1]Ph.D. Scholar, International Agribusiness Management Institute, Anand Agricultural University, Anand - 388110

[2]Ph.D. Scholar, Department of Agricultural Economics, B.A College of Agriculture, Anand Agricultural University, Anand - 388110

Abstract

This chapter presents a thorough examination of the impact of international trade and globalization on the agribusiness sector. It examines the significant impact of agribusiness on the global economy, explores the drivers and challenges associated with international trade and discusses the implications of globalization on various aspects of agricultural production, trade and market expansion. To begin, this chapter defines agribusiness as a comprehensive term that encompasses a range of activities within the food and agricultural system including production, processing, distribution and marketing. It emphasizes the economic importance of the sector, its potential for creating jobs and its role as a major source of exports in many countries. Moving forward, this chapter delves into the reasons why international trade and globalization are vital for the success of agribusiness. It highlights the growing global population and the need to increase food production to meet the demands of an estimated 9.7 billion people by 2050. International trade enables agricultural products to reach consumers worldwide, thereby creating more sales opportunities and expanding markets. This chapter acknowledges that globalization brings about both positive and negative effects on agribusiness. On the positive side, it facilitates market expansion, provides access to new customers and encourages the adoption of innovative technologies. However, it also poses challenges such as heightened competition, price volatility and the necessity to comply with diverse regulatory frameworks. In conclusion, this chapter discusses the future prospects and challenges in international agribusiness. It highlights emerging trends like evolving consumer preferences, advancements in technology, trade policies and the need for resilient supply chains. This chapter emphasizes the importance of collaboration and partnerships among stakeholders in order to navigate these challenges successfully and capitalize on future opportunities.

Keywords: International trade, globalization, agribusiness, market expansion, sustainable practices

Introduction

Agribusiness is a broad term that encompasses all aspects of the food and agricultural system from production to processing, distribution and marketing. It includes businesses that provide inputs to farmers such as seeds, fertilizers and equipment; businesses that process and distribute food products; and businesses that market food products to consumers. The agribusiness industry serves as a significant catalyst for economic development and the generation of employment opportunities in numerous countries. In the United States, for example, the agribusiness sector accounts for over 10% of GDP and employs over 20 million people (USDA, 2021). The sector is also a major source of exports with the United States exporting over $100 billion in agricultural products each year (USDA, 2021). Agribusiness is important to the global economy for a number of reasons. First, it provides food and fibre for a growing population. With the projected global population reaching 9.7 billion by 2050, there will be a heightened demand for food production to cater to this exponential growth. The agribusiness sector, in response, will play a pivotal role in meeting these increased food requirements. Additionally, agribusiness serves as a significant generator of employment opportunities, providing livelihoods for millions of individuals worldwide and serving as a key economic driver for rural communities. Third, agribusiness contributes to economic growth. The sector generates billions of dollars in revenue each year and it helps to drive economic development in many countries. The agribusiness sector encounters various obstacles, including climate change, food safety concerns, and competition from emerging markets. Nonetheless, the industry is actively engaged in innovation and the advancement of new technologies to tackle these challenges. For instance, agribusiness has embraced the development of novel agricultural techniques, such as genetically modified crops to enhance crop yields and elevate food quality. These innovations are crucial in overcoming the hurdles faced by the sector and ensuring its continued growth and sustainability. The development of new methods of food processing and distribution has helped to reduce food waste and improve food safety and the development of new marketing strategies has helped to increase the demand for agricultural products around the world. The agribusiness sector is a vital part of the global economy. It provides food and fibre for a growing population, creates jobs and contributes to economic growth. The sector faces a number of challenges but it is also innovating and developing new technologies to address these challenges. The agribusiness sector is well-positioned to play a major role in the global economy in the years to come.

India possesses vast agricultural lands with high crop yield potential, positioning it as a potential leader in the agricultural sector. The agricultural industry plays a significant role in the country's economy by generating employment opportunities for a large portion of the population accounting for 58.2% of total employment. To drive the sector's growth, a shift from subsistence farming to commercial farming is necessary. This shift entails implementing various changes, such as diversifying activities to enhance productivity, promoting value-added agriculture, implementing market reforms and creating new capacities in agro services. These measures aim to enable farmers to benefit from advanced technologies scientific agricultural practices and ultimately contributing to their success and the overall development of the agricultural sector in India. (Khan *et al* 2012).

Main features of new economic policy and its impact on agriculture

In the early 1990s, India embarked on significant economic policy reforms, driven by the need for macroeconomic stabilization and structural adjustments as advocated by international organizations like the World Bank and the International Monetary Fund. These policy changes aimed to address foreign exchange market imbalances through short-term measures such as reducing demand, devaluing the rupee, trade policy reforms, fiscal deficit reduction and removing barriers to foreign capital flows. Additionally, medium-term structural adjustment programs (SAPs) were introduced which encompassed fiscal, exchange rate, trade, and industrial policy reforms as well as changes in the public sector, financial sector and capital market. The overarching goal of these policy reforms was to realign domestic demand with available resources, bring about changes in the supply and production structure and eliminate external imbalances. The reforms also focused on liberalizing the economy by removing unnecessary controls and gradually integrating it with the global economy through the dismantling of tariff barriers. (Bhalla, 1995)

The history of international trade in agricultural products

The history of international trade in agricultural products dates back thousands of years, highlighting the significance of food as a valuable trade commodity and the desire for regions to access a diverse range of agricultural goods. Ancient civilizations engaged in agricultural trade such as the Sumerians in Mesopotamia (modern-day Iraq) and the ancient Egyptians along the Nile who exchanged grains, dates, and textiles. The middle Ages witnessed the trade of agricultural goods along established routes like the Silk Road and the Spice Trade including spices, herbs, silk, tea, coffee and sugar. The era of European exploration and colonization from the 15th to 19th centuries brought substantial changes to global agricultural trade. European powers established colonies in the Americas, Africa and Asia, cultivating crops like tobacco, sugar, cotton, coffee, and cocoa for export. This period also involved forced labour and the slave trade with enslaved people brought from Africa to work on plantations. The 18th-century agricultural revolution led to increased agricultural productivity through innovations in farming techniques and machinery, resulting in surplus production traded internationally, particularly in grains. The onset of industrialization in the 19th century transformed agricultural trade as industrialized nations required raw materials, increasing the demand for agricultural commodities from colonies and trading partners. The post-World War II period brought significant changes to international agricultural trade with the establishment of organizations like GATT and WTO, aiming to promote free trade and reduce barriers, including those related to agriculture. This period also saw debates surrounding agricultural subsidies, protectionist measures, and fair trade practices. The late 20th and 21st centuries witnessed the intensification of global agricultural trade and the integration of agribusiness into global supply chains. Technological advancements, improved transportation and communication systems have facilitated efficient cross-border movement of agricultural products. The rise of multinational agribusiness corporations has further influenced the global agricultural trade landscape.

The Effects of Globalization on Agribusiness

Globalization has had significant effects on agribusiness, shaping its structure, operations and impact on various stakeholders. Here are some key effects of globalization on agribusiness:

a. Market Expansion: Globalization has opened up new markets and expanded the customer base for agribusinesses. Increased international trade and improved market access have allowed agricultural products to reach consumers in different parts of the world. This has created opportunities for agribusinesses to increase sales, diversify their product offerings and cater to the preferences of global consumers.

b. Integration of Supply Chains: Globalization has led to the integration of agricultural supply chains on a global scale. Agribusinesses now operate across multiple countries, sourcing inputs, processing products and distributing them to various markets. This integration has resulted in the development of complex supply chain networks with increased coordination and collaboration among different actors in the agribusiness value chain.

c. Technological Advancements: Globalization has facilitated the transfer of technology and knowledge in agribusiness. Developed countries with advanced agricultural technologies have shared their expertise with developing nations, leading to the adoption of modern farming techniques, improved crop varieties and enhanced productivity. Technological advancements, such as precision agriculture, genetic engineering and digital tools, have enabled agribusinesses to improve efficiency, optimize resource use and enhance product quality.

d. Trade Liberalization: Globalization has been accompanied by efforts to reduce trade barriers and promote trade liberalization in agriculture. Trade agreements and organizations like the World Trade Organization (WTO) have aimed to lower tariffs, eliminate subsidies, and harmonize regulations and facilitating international trade in agricultural products. This has enabled agribusinesses to access larger markets, benefit from comparative advantages and compete globally.

e. Increased Competition: Globalization has intensified competition in the agribusiness sector. Agribusinesses now face competition from both domestic and international players. The removal of trade barriers has exposed domestic producers to global competition, necessitating improvements in efficiency and competitiveness. Agribusinesses have had to innovate, invest in research and development, and adopt advanced technologies to stay competitive in the global marketplace.

f. Environmental and Sustainability Concerns: The globalization of agribusiness has raised environmental and sustainability concerns. Increased agricultural production and trade have led to deforestation, biodiversity loss, water pollution and greenhouse gas emissions. Agribusinesses are under pressure to adopt sustainable practices, minimize their environmental footprint, address social and ethical issues in their operations.

g. Impact on Small-scale Farmers: Globalization has mixed effects on small-scale farmers. While some small-scale farmers have benefited from increased market access and export opportunities, others have faced challenges due to competition from large-scale agribusinesses, price fluctuations and limited access to resources and markets. Ensuring inclusive growth and supporting the sustainability of small-scale farming communities is a key challenge in the face of globalization.

h. Food Safety and Quality Standards: Globalization has led to the harmonization and standardization of food safety and quality regulations. Agribusinesses need to comply with various international standards to access global markets. This has improved consumer confidence in agricultural products but compliance with standards can pose challenges for small-scale farmers and producers with limited resources.

i. Social and Cultural Impacts: Globalization has influenced the social and cultural aspects of agribusiness. The spread of global food chains and Westernized diets has affected traditional food systems and local agricultural practices. Agribusinesses have also faced scrutiny regarding labour practices, fair trade and the impact of their operations on local communities and indigenous populations.

These effects of globalization on agribusiness highlight both opportunities and challenges in the global agricultural sector. Agribusinesses must navigate these changes while addressing sustainability, inclusivity and the evolving needs of consumers in an interconnected world.

The impact of globalization and reform on the Indian informal sector

The impact of globalization and economic reforms is primarily experienced by the formal sector while the informal sector operates outside the scope of regulations. However, economic reforms can still affect the informal sector in several ways. One such impact is observed through trade liberalization which often leads to reduced prices of tradable goods. If the products of the informal sector are close substitutes for tradable goods and services, they may experience similar price reductions. Consequently, trade liberalization can affect the prices and competitiveness of informal-sector products. Furthermore, trade liberalization can indirectly affect the informal sector through labour market linkages. When individuals lose their jobs in the formal sector due to economic reforms, they may turn to the informal sector for alternative employment opportunities. This increased labour supply to the informal sector can lead to lower incomes for informal workers. This relationship between the formal and informal sectors is often considered in models of rural-urban migration and segmented labour markets. It is important to recognize that there are other connections and interdependencies between the formal and informal sectors which vary in significance depending on the specific context. These linkages can have implications for the impact of economic reforms on the informal sector. Therefore, when assessing the effects of reforms on the informal sector, it is crucial to empirically evaluate the most relevant linkages and their specific consequences (Siggel, E. 2010).

Positive effects of globalization on agribusiness

1. Market Expansion: Globalization opens up new markets and increases the customer base for agribusinesses. It allows agricultural products to reach consumers in different parts of the world, leading to increased sales and market opportunities.

   a. Increased Sales and Revenue: By expanding into new markets, agribusinesses can increase their sales and revenue streams. This is particularly crucial in a globalized economy where businesses need to explore diverse customer bases to maintain growth and profitability.

   b. Diversification: Market expansion allows agribusinesses to diversify their customer base and reduce reliance on a single market or product. By accessing multiple markets, agribusinesses can mitigate risks associated with market volatility, changes in consumer preferences and other uncertainties.

   c. Access to New Customers: Market expansion provides agribusinesses with access to new customers who may have different preferences and demands. This allows agribusinesses to adapt their products, packaging, branding and marketing strategies to cater to the specific needs of different markets, thereby enhancing customer satisfaction and loyalty.

   d. Competitive Advantage: Expanding into new markets can give agribusinesses a competitive advantage over their competitors. By reaching untapped markets or penetrating existing markets more effectively, agribusinesses can secure a larger market share, increase their brand visibility and strengthen their position in the industry.

   e. Growth Opportunities: Market expansion presents growth opportunities for agribusinesses. It allows them to capitalize on emerging trends such as the increasing demand for organic or specialty products and leverage their unique strengths to capture new market segments.

   f. Knowledge and Learning: Entering new markets exposes agribusinesses to different business environments, cultural dynamics and regulatory frameworks. This provides valuable learning experiences and insights that can enhance their overall business knowledge and capabilities.

2. Access to Technology and Innovation: Globalization facilitates the transfer of technology and knowledge in agribusiness. Developing countries can adopt advanced agricultural technologies, leading to improved productivity, efficiency and competitiveness.

3. Economies of Scale: Globalization allows agribusinesses to operate at a larger scale, leading to cost savings through economies of scale. This can improve efficiency, reduce production costs and enhance profitability.

4. Knowledge Exchange and Collaboration: Globalization promotes knowledge exchange and collaboration among agribusinesses across different regions. This facilitates learning, innovation and the adoption of best practices, leading to improved agricultural practices and productivity gains.

5. Diversification and Risk Management: Globalization allows agribusinesses to diversify their markets and reduce dependence on a single market or product. This helps mitigate risks associated with market fluctuations, changes in consumer preferences and climate-related challenges.

Negative effects of globalization on agribusiness

a. Displacement of Small-Scale Farmers: Globalization can lead to the displacement of small-scale farmers, particularly in developing countries. Agribusinesses with access to large-scale production methods and global distribution networks may outcompete small farmers leading to rural unemployment and socio-economic disparities. (Patel, R. (2018).

b. Environmental Degradation: The globalization of agribusiness can contribute to environmental degradation. Increased production and trade in agricultural products often involve unsustainable practices such as deforestation, soil degradation, water pollution and greenhouse gas emissions, which can harm ecosystems and biodiversity.

c. Loss of Cultural Diversity: Globalization can lead to the homogenization of agricultural practices and the loss of traditional farming systems and cultural diversity. The adoption of standardized production methods and the dominance of global agribusinesses can erode local knowledge and agricultural traditions.

d. Dependence on Inputs and External Markets: Globalization can create a dependency on external inputs, such as seeds, fertilizers and pesticides, which are often controlled by multinational corporations. Agribusinesses may become reliant on these inputs, leading to increased production costs and potential health and environmental risks.

The impact of globalization on agricultural production, prices, and trade.

1. Agricultural Production:

a. Technological Advancements: Globalization has facilitated the transfer of agricultural technologies and knowledge across borders. This has led to the adoption of advanced farming practices, improved seeds, precision agriculture techniques and efficient use of inputs. Technological advancements have increased productivity, crop yields and quality of agricultural produce.

b. Access to Inputs: Globalization has enhanced access to agricultural inputs such as fertilizers, pesticides, machinery and farm equipment. Improved availability and affordability of inputs have supported agricultural production by enabling farmers to achieve higher yields and better crop protection.

c. Specialization and Comparative Advantage: Globalization has encouraged countries to specialize in the production of agricultural commodities that they have a comparative advantage in. This has led to the efficient allocation of resources and increased production. Regions or countries with favourable climate conditions or specific expertise in certain crops have capitalized on their strengths to become major exporters in global agricultural markets.

d. Diversification of Production: Globalization has facilitated the diversification of agricultural production. Farmers can now grow a wider range of crops and agricultural products, responding to changing consumer demands and preferences both domestically and internationally. This diversification has led to expanded product portfolios and increased market opportunities for agricultural producers.

e. Integration of Value Chains: Globalization has resulted in the integration of agricultural value chains on a global scale. Producers, processors, distributors and retailers are increasingly interconnected across countries, enabling a seamless flow of agricultural products from farm to consumer. This integration has enhanced efficiency, improved supply chain management and reduced transaction costs.

f. Market Expansion: Globalization has opened up new markets and increased the customer base for agribusinesses. It allows agricultural products to reach consumers in different parts of the world, leading to increased sales and market opportunities. Agribusinesses can tap into international markets, diversify their customer base and reduce dependence on local markets (Vollrath, T. L.2018).

g. Adoption of Sustainable Practices: Globalization has brought greater awareness, scrutiny of environmental and social impacts of agricultural production. As a result, there has been an increasing focus on sustainable agriculture practices such as organic farming, agro ecology and fair trade. Globalization has provided opportunities for the dissemination and adoption of these practices, leading to more environmentally friendly and socially responsible agricultural production.

h. Challenges and Disparities: While globalization has brought numerous benefits, it has also presented challenges and disparities in agricultural production. Small-scale farmers and marginalized regions may face difficulties in competing with larger agribusinesses, particularly due to pricing pressures and economies of scale. Additionally, global trade rules and subsidies can affect the competitiveness of certain agricultural sectors, creating imbalances and inequalities in the global market.

2. Prices:

a. Market Integration: Globalization has increased the integration of agricultural markets, connecting producers and consumers across different regions and countries. This integration has the potential to lead to greater price transparency and more efficient price discovery mechanisms, as market participants can access and respond to global price signals. It can also promote competition, which can influence agricultural prices.

b. Price Volatility: Globalization has exposed agricultural commodities to greater price volatility. Fluctuations in global supply and demand, weather conditions, changes in trade policies and other factors can have significant impacts on agricultural prices. Rapid dissemination of information and increased interconnectedness through globalization can amplify price movements, creating both opportunities and challenges for agricultural producers and consumers.

c. International Trade: Globalization has facilitated increased international trade in agricultural products. Reductions in trade barriers, such as tariffs and quotas, can enhance market access and expand export opportunities for agricultural producers. Increased trade can influence prices by affecting the supply and demand dynamics in domestic and international markets. Changes in trade policies, currency exchange rates and global trade agreements can also impact agricultural prices.

d. Competition and Market Power: Globalization can intensify competition in agricultural markets. Agribusinesses and farmers may face competition from both domestic and international producers, which can influence prices. Market power and concentration in the agribusiness sector can also impact prices, as large multinational corporations may have the ability to influence supply chains and exert pricing pressure.

e. Input Costs: Globalization can affect the costs of agricultural inputs, such as fertilizers, pesticides, machinery and labour. Access to global markets can provide opportunities for farmers to obtain inputs at lower costs. However, factors such as fluctuations in commodity prices, transportation costs and exchange rate fluctuations can also influence input costs, which, in turn, can affect agricultural prices.

f. Government Policies: Government policies, including agricultural subsidies, trade restrictions and domestic support programs, can have a significant impact on agricultural prices in the context of globalization. These policies can influence domestic production levels, market competitiveness and price stabilization efforts.

g. Price Compression: Globalization can lead to price compression in certain agricultural commodities. Increased competition and market integration may result in price convergence, as prices tend to align across different markets. This can put pressure on agricultural producers, particularly in regions with higher production costs, as they may face challenges in competing with lower-priced imports.

3. Trade:

a. Market Expansion: Globalization has opened up new markets and increased the scope for agricultural trade. Reductions in trade barriers, such as tariffs and quotas, have facilitated the flow of agricultural products across borders, enabling producers to access larger and more diverse markets. This market expansion has created opportunities for increased export revenues and market diversification for agribusinesses and farmers.

b. Trade Liberalization: Globalization has led to the liberalization of agricultural trade through various international agreements and organizations, such as the World Trade Organization (WTO). Trade liberalization has aimed to reduce trade barriers, promote fair trade practices and provide a level playing field for agricultural producers and exporters. This has encouraged greater participation in global agricultural trade and increased competition.

c. Export Opportunities: Globalization has enabled agricultural producers to tap into international markets and expand their export opportunities. Producers can reach consumers in different parts of the world, thereby increasing sales and revenue potential. This has been particularly beneficial for countries with a comparative advantage in specific agricultural products or regions with high-quality specialized products.

d. Import Access: Globalization has facilitated access to a wider range of agricultural products through imports. Importing countries can access agricultural products that are not locally produced or are more cost-effective to import. This allows for greater consumer choice and availability of diverse agricultural products throughout the year.

e. Supply Chain Integration: Globalization has facilitated the integration of agricultural supply chains on a global scale. Producers, processors, distributors, and retailers are interconnected across countries, enabling seamless movement of agricultural products from farm to consumer. This integration has improved supply chain efficiency, reduced transaction costs, enhance the timeliness and reliability of agricultural trade.

f. Technology Transfer: Globalization has facilitated the transfer of agricultural technologies, knowledge and best practices across borders. This has resulted in the adoption of advanced farming techniques, improved seeds, precision agriculture methods and efficient supply chain management. Technological advancements have enhanced agricultural productivity and quality, contributing to increased agricultural trade.

g. Trade Disputes and Challenges: Globalization has also brought about challenges and trade disputes in agricultural trade. Disputes can arise over issues such as market access, trade barriers, subsidies and sanitary and phytosanitary standards. These disputes can impact the flow of agricultural trade and create uncertainties for producers and exporters.

h. Impact on Small-Scale Farmers: The impact of globalization on agricultural trade can be uneven, with small-scale farmers facing challenges in competing with larger agribusinesses. Limited resources, lower economies of scale, lack of market information and infrastructure can hinder the ability of small-scale farmers to participate in global agricultural trade.

Trade agreements impact agribusiness in various ways, here are some specific agreements that have significant implications for the agricultural sector:

i. **Agreement on Agriculture (AoA)** - This agreement is part of the World Trade Organization (WTO) framework and aims to reform trade in agricultural products. The AoA establishes rules and commitments related to market access, domestic support and export subsidies. It has influenced agricultural trade policies and practices globally, affecting agribusinesses' market opportunities and competitiveness.

ii. **United States-Mexico-Canada Agreement (USMCA)** - This trade agreement replaces NAFTA and has provisions specific to agribusiness. It includes provisions for improved market access for agricultural products, including dairy, poultry and wheat. The USMCA also addresses issues related to biotechnology and intellectual property rights in the agricultural sector.

iii. **European Union Common Agricultural Policy (CAP)** - The CAP is a significant policy framework within the European Union (EU) that impacts agribusiness. It sets rules and regulations for agricultural subsidies, market interventions and trade policies within the EU. The CAP influences agricultural production, prices and market access for agribusinesses operating within the EU.

iv. **Comprehensive and Progressive Agreement for Trans-Pacific Partnership (CPTPP)** - The CPTPP is a trade agreement among 11 Pacific Rim countries, including Australia, Canada, Japan, and Vietnam. It includes provisions that specifically impact the agricultural sector, such as tariff reductions, improved market access for agricultural products and commitments to promote sustainable agricultural practices.

v. **Mercosur** - Mercosur is a regional trade agreement among South American countries, including Argentina, Brazil, Paraguay and Uruguay. The agreement has provisions related to agricultural trade and cooperation, including tariff reductions and market access commitments. Agribusinesses within Mercosur member countries benefit from preferential trade arrangements and enhanced regional integration.

Future Prospects and Challenges in International Agribusiness

a. International agribusiness faces a range of prospects and challenges as the world's population continues to grow, consumer preferences evolve and global trade dynamics change. Here are some key future prospects and challenges in international agribusiness:

b. Increasing Global Demand: The global population is projected to reach 9.7 billion by 2050, leading to increased demand for food, feed, fibre and fuel. This presents significant opportunities for agribusinesses to expand their production and supply capabilities to meet the growing needs. However, meeting this demand sustainably while ensuring food security, reducing environmental impacts, addressing social and economic inequalities will be a major challenge.

c. Evolving Consumer Preferences: Consumer preferences are changing, driven by factors such as health consciousness, environmental concerns and ethical considerations. Consumers are increasingly seeking out organic, locally sourced and sustainably produced agricultural products. Agribusinesses will need to adapt to these evolving preferences by adopting sustainable and responsible production practices, transparent supply chains and effective communication strategies to build consumer trust and loyalty.

d. Technological Advancements: Advancements in technology, such as precision agriculture, automation and digitalization, hold immense potential for improving productivity, efficiency and sustainability in agribusiness. These technologies enable better resource management, data-driven decision-making and enhanced traceability throughout the supply chain. Agribusinesses that embrace and invest in innovative technologies will have a competitive edge in the global market.

e. Climate Change and Resource Constraints: Climate change poses significant challenges to agribusiness by impacting crop yields, water availability and overall agricultural productivity. Adapting to changing climatic conditions and mitigating greenhouse gas emissions will be crucial for long-term sustainability. Additionally, resource constraints, such as water scarcity and land degradation will require innovative approaches to optimize resource use and minimize environmental impacts.

f. Trade Policies and Geopolitical Dynamics: Trade policies and geopolitical factors play a significant role in shaping international agribusiness. Trade agreements, tariffs, subsidies and non-tariff barriers impact market access, pricing and competitiveness. Agribusinesses must navigate these complex trade environments, understand market dynamics and adapt their strategies accordingly to seize opportunities and mitigate risks.

g. Supply Chain Resilience and Traceability: Recent disruptions, such as the COVID-19 pandemic and climate-related events have highlighted the need for resilient and transparent supply chains in agribusiness. Ensuring supply chain resilience, diversification and traceability will be essential to mitigate risks and maintain uninterrupted trade flows. Agribusinesses will need to invest in technologies, partnerships and risk management strategies to enhance supply chain resilience.

h. Regulatory and Sustainability Requirements: Increasingly stringent regulations and sustainability requirements, both at the national and international levels, are shaping the agribusiness landscape. Compliance with food safety standards, environmental regulations and sustainability certifications is becoming critical for accessing markets and maintaining consumer trust. Agribusinesses must stay updated with evolving regulations and invest in sustainable practices to remain competitive and meet market expectations.

i. Social and Ethical Considerations: Agribusinesses are increasingly being scrutinized for their social and ethical practices, including labor rights, animal welfare and fair trade. Stakeholders, including consumers, investors and advocacy groups are demanding greater transparency and accountability throughout the supply chain. Agribusinesses that prioritize social responsibility, ethical sourcing and inclusivity will have a competitive advantage in the market.

j. Collaboration and Partnerships: Addressing the complex challenges facing international agribusiness often requires collaboration and partnerships across various stakeholders, including governments, research institutions, NGOs and local communities. Collaborative initiatives can facilitate knowledge-sharing, innovation and capacity-building, leading to more sustainable and inclusive agribusiness practices.

Conclusion

International trade and globalization had a profound impact on agribusiness. Agribusiness, which encompasses all aspects of the food and agricultural system, has benefited from increased market access, expanded customer base and the ability to tap into global supply chains. The globalization of agribusiness has facilitated the exchange of agricultural products across borders, leading to economic growth, job creation and improved food security. Globalization has opened up new markets for agribusinesses, allowing them to reach consumers in different parts of the world. This has led to increased sales, market opportunities and the potential for higher profits. Agribusinesses have also benefited from advancements in transportation and communication technologies, which have made it easier to connect with international markets and manage global operations. Furthermore, globalization has contributed to the development and adoption of innovative agricultural practices and technologies. This has resulted in increased productivity, improved efficiency and the ability to meet the growing demands of a rapidly expanding global population. Agribusinesses have embraced sustainable practices, such as organic farming, precision agriculture and agroecology to address environmental concerns and consumer preferences for sustainable products. However, globalization has also presented challenges for agribusinesses. The intensification of global competition, fluctuating commodity prices and the volatility of international markets have created uncertainties and risks. Additionally, issues such as climate change, resource constraints and changing regulatory landscapes have posed challenges for sustainable agricultural production and trade.In navigating the complexities of international trade and globalization, agribusinesses must be adaptable, innovative and responsive to evolving market dynamics. They need to embrace sustainable practices, invest in technology, build resilient supply chains, and comply with international standards and regulations. Collaboration and partnerships with various stakeholders, including governments, research institutions and local communities are also crucial for addressing global challenges and achieving sustainable development in agribusiness. Overall, international trade and globalization have provided significant opportunities for agribusinesses to expand their markets, increase profitability and contribute to global food security. However, it is essential to address the challenges and ensure that the benefits of globalization are shared equitably and sustainably, considering social, economic, and environmental aspects. By doing so, agribusinesses can play a vital role in feeding the world's growing population while preserving the planet's resources for future generations.

References

Asif Khan, D. M., Abbasi, D. E. H., & KM, N. (2012). Indian Agribusiness and Globalisation: Prospects and Challenges. *International Journal of Management Research and Development (IJMRD)*, *2*(1). 70-83.

Bhalla, G. S. (1995). Presidential Address: Globalisation and Agricultural Policy in India. *Indian Journal of Agricultural Economics*, *50*(902-2018-3344), 7-26.

Patel, R. (2018). The long green revolution. Journal of Peasant Studies, 45(1), 1-33.

Siggel, E. (2010). The Indian informal sector: The impact of globalization and reform. *International Labour Review*, 149(1), 93–105.

United States Department of Agriculture. (2021). Agricultural Economic Insights. Retrieved from https://www.ers.usda.gov/topics/farm-economy/agricultural-economic-insights/

United States Department of Agriculture. (2021). Agricultural Trade. Retrieved from https://www.ers.usda.gov/topics/international-markets-trade/agricultural-trade/

Vollrath, T. L. (2018). The Effects of Globalization on Market Expansion in Agribusiness. Journal of Agricultural Economics, 70(2), 456-472.

13

Quality Assurance and Food Safety in Agribusiness

Vaishnavi[1*], Aravinda, B. J.[2] and Raghavendra, P. K.[3]

[1*]Ph.D. Scholar, Department of Agricultural Economics, University of Agricultural Sciences, Bangalore-65

[2]Ph.D. Scholar, Institute of Agri-Business Management, University of Agricultural Sciences, Bangalore-65

[3]Ph.D. Scholar, Institute of Agri-Business Management, University of Agricultural Sciences, Bangalore-65

(*Corresponding author Mail Id: vaishnavimadari@hotmail.com

Abstract

Food safety and nutrition are closely interconnected, as unsafe food can result in illness and inadequate nourishment, impacting vulnerable populations. Collaborative endeavors involving governments, producers, suppliers, distributors, and consumers are pivotal in guaranteeing food safety in the modern era, considering the intricate nature of global food supply chains. Food safety entails employing scientific methods in handling, preparing, and storing food to prevent instances of foodborne diseases. Quality control initiatives are implemented worldwide, particularly for animal-derived products, to safeguard human well-being. Upholding stringent quality standards in agribusiness is indispensable for ensuring consumer welfare, regulatory adherence, brand reputation, risk management, and market entry. This endeavor contributes to the overall prosperity of society, environmental sustainability, and economic implications. By giving prominence to quality assurance and food safety, agribusinesses can heighten their competitiveness, safeguard consumers, and foster a responsible and sustainable food industry.

Keywords: Quality Assurance, Food Safety, Good Agricultural Practices (GAP), Good Manufacturing Practice (GMP), Hazard Analysis Critical Control Point (HACCP), International Standard Organization (ISO), Total Quality Management (TQM).

Introduction

Indian economy is still agrarian, because about 60 per cent population is directly dependent upon agriculture. Agriculture forms the foundation of agribusiness and encompasses the cultivation of crops, livestock rearing, and other related activities. It plays a crucial role in providing a sustainable and abundant supply of food, feed, fiber, and fuel for the growing global population. Modern agricultural practices involve the use of technology, scientific research, and best management practices to optimize productivity, enhance crop yields, and promote sustainable farming methods.

Food is any substance, whether processed, semi processed or raw, which is intended for human consumption including drink and other substances which has been used in its manufacturer, preparation or treatment. Safety and quality of food are increasingly prominent worldwide, especially in relation to the well-being of individuals. Consequently, numerous nations have implemented programs to monitor the quality of all food constituents, specifically those derived from animals. Food safety constitutes a scientific field that encompasses the proper management, preparation, and preservation of food to mitigate the risk of foodborne diseases. This necessitates adherence to a series of regular procedures aimed at averting the occurrence of potentially grave health risks (Feyera Gemeda, 2023)

Food safety:

Food safety refers to the measures and procedures implemented to ensure that food products are safe for consumption and free from contaminants, harmful substances, and pathogens that could cause illness or harm to consumers. Maintaining food safety throughout the entire food supply chain is essential to protect public health, prevent foodborne diseases, and promote consumer confidence.

Quality assurance:

The ISO definition, "the assembly of all planned and systematic actions necessary to provide adequate confidence that a product, process, or service will satisfy given quality requirements." Quality Assurance, in contrast to quality control, is the implementation of quality checks and procedures to immediately correct any failure and mistake that is able to reduce the quality of the interim products at every production step. Thus, the desired high quality of the final product is planned and obtained.

Hazard:

A hazard is a biological, chemical, or physical element that, if left uncontrolled, significantly increases the risk of causing illness or injury. Routine practices such as maintaining personal hygiene, employing appropriate food processing techniques, applying high-temperature heat treatment, ensuring thorough cooking, and avoiding temperature conditions conducive to bacterial growth, help control a diverse range of foodborne illnesses (Addis and Sisay, 2015).

Therefore, food regulation is prioritized as a compulsory recurring practice implemented by national or local governing bodies to safeguard consumers and guarantee the safety, quality, and suitability of all food items throughout their production, handling, storage, processing, and distribution for human consumption. Before a given food item is consumed, it should conform to safety and quality requirements, honestly and accurately labeled as prescribed by law (CFSAN, 2007). Factors that contribute to potential food hazards encompass inadequate farming methods coupled with conventional dairy production, substandard hygiene practices throughout the entire food supply chain, absence of preventive and control measures in food processing and preparation procedures, improper use of chemicals, contaminated raw materials, ingredients, and water sources, and insufficient or inappropriate storage conditions (Battu et al., 2004; Bergdoll, 2006). Hence, it necessitates the implementation of effective quality assurance and control methods to manage and mitigate these hazards.

Quality assurance and food safety are critical aspects encompassing various practices and measures to ensure the production, processing, and distribution of safe and high-quality agricultural products. The integration of agriculture and food safety is essential to protect consumers, promote public health, and maintain the integrity of the food supply chain.

Food Safety System

» **Traditional food safety system:** The conventional food safety system has often been criticized as unreliable. It follows a reactive approach, primarily relying on governmental oversight, conducting inspections and tests on final products, lacking structured risk analysis, and sometimes falling short in achieving satisfactory risk reduction levels (FAO, 1997). As a result of these factors, the traditional food safety system continues to be ineffective, unable to adequately address and resolve numerous existing issues, and incapable of effectively handling the wide array of intricate, persistent, and evolving challenges that impact various aspects of the food supply chain.

» **Science risk-based food safety system:** A science-based approach to food safety encompasses a range of activities, including good agricultural practices, good hygienic practices, good manufacturing practices, and the implementation of Hazard Analysis and Critical Control Points (HACCP). This approach has embraced the adoption of risk analysis as a comprehensive framework to systematically and scientifically address food safety issues, enhancing the quality of decision-making at every stage of the food chain (FAO, 2003).

India ranks second globally in food and agricultural production. Food production in India for the year 2020-21 showing the rankings and total production of various commodities (Table 1). As the Indian economy has diversified and grown, yet it is still the country's ensuring food safety and quality assurance is crucial, given the large quantities produced. Robust measures are needed to prevent contamination, maintain nutritional value, and mitigate health risks. Regular inspections, testing, and traceability systems help ensure safety, maintain consumer trust, and meet regulatory requirements. Strengthening food safety systems through infrastructure investments, training programs, and promoting awareness is essential to meet growing demands and enhance India's reputation in domestic and international markets.

Table 1: Food Production in India (2020-21)

Commodity	Rank	Total production
Cereals and grains	2	323.55 million tonnes
Milk	1	330 million metric tonnes
Meat	8	8.80 million tonnes
Poultry	3	114.58 bn eggs
Fruits and Vegetables	2	311.7 million tonnes
Marine	2	14.16 million metric tonnes

Source: www.faostat.com

Major Challenges in Quality Assurance and Food Safety

- » Globalization of the Food Supply Chain: Complex supply chains make it challenging to monitor and control food quality and safety across multiple countries and continents
- » Food Fraud and Counterfeiting: Misrepresentation of food products through adulteration, substitution, and mislabeling poses a significant threat to quality assurance
- » Changing Consumer Expectations: Consumers demand transparency, traceability, and higher standards for food safety, presenting a challenge for the industry
- » Emerging Foodborne Pathogens: Ongoing evolution and emergence of new pathogens require robust surveillance and control measures to ensure food safety
- » Regulatory Compliance: Keeping up with evolving food safety regulations and implementing new requirements across different regions is a complex and resource-intensive task
- » Microbiological Safety: Foodborne illnesses are primarily caused by bacteria such as Salmonella, Campylobacter, and Listeria, highlighting the importance of preventing microbial contamination
- » Chemical Safety: Contaminants and non-food grade additives in food, including pesticide residues and heavy metals, pose risks to consumer health
- » Personal Hygiene: Poor personal hygiene practices among food handlers can lead to the transmission of pathogens and foodborne illnesses
- » Environmental Hygiene: Inadequate waste disposal and unsanitary conditions in food processing areas contribute to food contamination and spoilage risks.

Why is Safe Food Supply Important?

Ensuring a secure food supply is crucial due to the substantial impact it has on both the health and economic well-being of societies and nations. Globally, it has been approximated that over one billion cases of diarrheal illness caused by food poisoning transpire each year, resulting in the tragic loss of approximately 3 million children annually, primarily in less developed areas (www.fao.org).

From an economic standpoint, ensuring adequate access to safe and nutritious food is vital for sustaining life, fostering good health, and driving economic growth. In a study, the cost-of-illness model was expanded to incorporate economic evaluations encompassing medical expenses, productivity decline, and mortality associated with foodborne illnesses (Fred Fung, 2018).

Food Safety And Quality Assurance (FSQA) System Elements

International inter-governmental organizations have been evolving approaches with shared responsibilities for governments, industry and consumer to ensure safe and quality food (Fig. 1). The approach aims that the end food product should be entirely safe and of high quality. Specific safety and quality standards: In FSQA system, Codex specifies maximum permissible residue limits of important chemical contaminants in food; high microbiological quality; important quality attributes; maximum limits of use of additives, and labelling and packaging requirements. Minimizing contaminants: An important aspect of Codex standards is specification of maximum permissible limits for residues of several toxic chemical contaminants.

Why are they important?

Safeguarding the health of a nation's citizens is a fundamental responsibility of the government, intricately tied to the pursuit of various Sustainable Development Goals (SDGs), notably SDG 2 and 3. Simultaneously, food control plays a pivotal role in ensuring equitable practices within the food trade, fostering economic opportunities for all stakeholders involved throughout the food chain.

In an era characterized by rapid advancements in food technologies and a continuously expanding global food trade, the control of foodborne hazards throughout the entire food chain has assumed a critical significance. To effectively address the complex emerging challenges of the 21st century, food control systems must remain current with the latest advancements, adhere to the principles of risk analysis, and align with international standards and best practices established by Codex Alimentarius. The international food trade, a thriving industry valued at 2 trillion dollars annually, involves the production, marketing, and transportation of billions of tonnes of food (FAO, 2022).

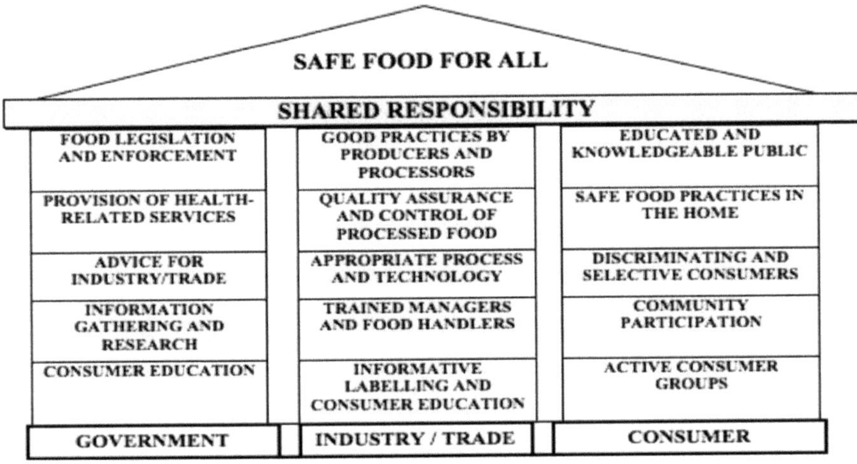

Fig. 1: Shared responsibility-safe food for all

Source: indiadairy.com

Importance of Quality Assurance and Food Safety in Agribusiness

Quality assurance and food safety are crucial aspects of the agribusiness industry. Quality assurance focuses on meeting or exceeding customer expectations through standardized processes and guidelines, while food safety involves preventing contamination and ensuring that food products are safe for consumption. Both aspects work together to ensure that agricultural products meet quality standards and are safe throughout the production and supply chain.

- Quality assurance and food safety protect consumer health by preventing harmful contaminants in food products
- Maintaining high-quality standards builds consumer trust and fosters brand reputation
- Compliance with regulations is crucial for market access and international trade
- Implementing quality control measures helps prevent costly product recalls
- Quality assurance enhances product quality and increases market competitiveness
- Adhering to food safety standards facilitates market access in the global agribusiness industry

- Consumer protection is ensured through quality assurance and food safety measures
- Brand reputation is strengthened by prioritizing quality and safety
- Supply chain management maintains consistent quality standards throughout the production process
- Risk mitigation measures proactively address potential risks, reducing foodborne illnesses and losses
- International trade requires meeting food safety and quality standards for global market access

Objectives of Quality Assurance in Agribusiness

- **Product Quality:** Maintain and improve the quality of agricultural products, ensuring they meet standards and regulations
- **Food Safety:** Ensure the safety of agricultural products, minimize contamination, and reduce the risk of foodborne illnesses
- **Process Efficiency:** Optimize agribusiness processes to minimize waste, reduce costs, and increase productivity
- **Traceability:** Establish systems to accurately track and document the movement of agricultural products throughout the supply chain
- **Compliance with Regulations:** Ensure adherence to relevant laws, regulations, and industry standards
- **Customer Satisfaction:** Meet or exceed customer expectations by consistently delivering high-quality products and improving based on customer feedback

Significance of Maintaining High-Quality Standards in Agricultural Industry

- **Consumer Health and Safety:** High-quality standards ensure that the food produced is safe for consumption. They encompass various aspects such as the proper use of fertilizers, pesticides, and veterinary drugs, adherence to hygiene and sanitation practices, and monitoring of potential contaminants. By maintaining these standards, the risk of foodborne illnesses, allergies, and other health issues is significantly reduced, safeguarding the well-being of consumers

- **Reputation and Consumer Confidence:** High-quality agricultural products contribute to the reputation and credibility of farmers, producers, and the industry as a whole. Consumers expect reliable and safe food choices, and meeting or exceeding quality standards builds trust. This trust encourages repeat purchases, enhances brand loyalty, and boosts the overall confidence of consumers in the agricultural industry
- **Market Access and Competitiveness:** International trade in agricultural products requires adherence to quality standards to ensure market access. Many countries have stringent import regulations and quality control measures in place to protect their domestic markets and consumers. Compliance with these standards allows agricultural producers to access global markets and compete on a level playing field, expanding their customer base and increasing profitability
- **Environmental Sustainability:** High-quality standards often incorporate environmentally friendly practices that promote sustainability. These may include the responsible use of water resources, soil conservation techniques, integrated pest management, and organic farming methods. By adhering to these standards, the agricultural industry can minimize its impact on ecosystems, preserve biodiversity, and contribute to a more sustainable future
- **Economic Impact:** The agricultural sector plays a crucial role in the economy of many countries. Maintaining high-quality standards can enhance economic growth and stability by ensuring consistent production, reducing post-harvest losses, and increasing the value of agricultural products. High-quality agricultural products also have better market value and can command premium prices, leading to increased income for farmers and stakeholders
- **Regulatory Compliance:** Governments and regulatory bodies implement quality standards to protect consumer interests, maintain public health, and ensure fair trade practices. Compliance with these standards is not only a legal requirement but also helps in avoiding penalties, legal issues, and reputational damage. By adhering to high-quality standards, the agricultural industry demonstrates its commitment to responsible practices and regulatory compliance.
- **Consumer Education:** Educating consumers about the importance of quality assurance in agribusiness is crucial. Providing information about

the production methods, certifications, and quality standards followed by agribusinesses helps consumers make informed choices. Consumer awareness and demand for high-quality, sustainable products can drive improvements in the industry.

Compliance Requirements for Agribusiness under Quality Assurance

» **Product Specification:** Establishing clear product specifications and standards is crucial for ensuring consistent quality in agribusiness products. Product specifications outline the desired characteristics, composition, and performance criteria of a product. These specifications serve as a reference point for both producers and consumers, enabling them to understand and communicate the desired quality attributes. By clearly defining product specifications, agribusinesses can set quality benchmarks that guide their production processes. This helps in achieving uniformity in product quality, meeting customer expectations, and complying with industry regulations and standards. Product specifications also aid in managing the supply chain effectively by providing suppliers with specific requirements for raw materials and components

» **Raw Material Evaluation:** Evaluating and selecting quality raw materials is a critical aspect of ensuring the overall quality of agribusiness products. The process involves assessing the physical, chemical, and biological attributes of raw materials to ensure they meet the required standards. Agribusinesses need to establish criteria for evaluating raw materials, such as freshness, purity, nutritional value, and absence of contaminants or harmful substances. This evaluation can involve inspections, sampling, and laboratory testing. By selecting high-quality raw materials, agribusinesses can enhance the chances of producing safe, nutritious, and desirable products

» **Process Control:** Process control measures play a vital role in maintaining consistent quality throughout the production cycle. These measures involve monitoring and adjusting various parameters during manufacturing, processing, and packaging stages to ensure adherence to predetermined quality standards. Process control may include activities like monitoring temperature, humidity, time, pressure, pH levels, and other relevant variables. By implementing robust process control measures, agribusinesses can minimize variations in product quality, reduce defects, optimize resource utilization, and improve overall efficiency. This helps in delivering products that consistently meet customer expectations and comply with regulatory requirements.

» **Testing and Analysis:** Laboratory testing and analysis are essential for verifying product quality in agribusiness. These activities involve analyzing samples of raw materials, intermediate products, and finished goods to assess their compliance with established specifications and standards. Laboratory testing can encompass various techniques, such as chemical analysis, microbiological testing, sensory evaluation, and physical measurements. Through these tests, agribusinesses can identify any deviations or non-conformities and take corrective actions to maintain product quality. Testing and analysis also provide valuable feedback for process improvement, product development, and meeting regulatory requirements

» **Documentation and Record-keeping:** Maintaining accurate records is crucial for traceability, accountability, and continuous improvement in agribusinesses. Documentation includes recording relevant information at every stage of the production process, such as raw material sources, production parameters, quality control activities, and test results. Accurate documentation enables agribusinesses to trace the origin of raw materials, track product flow, and identify potential issues in the production chain. It also helps in demonstrating compliance with regulatory standards, facilitating product recalls if necessary, and providing evidence of due diligence. Furthermore, documentation serves as a valuable resource for analysis, benchmarking, and making informed decisions to enhance product quality and efficiency.

Quality Assurance in Supply Chain Management

Quality assurance in supply chain management refers to the processes and activities implemented to ensure that products and services meet or exceed customer expectations and comply with relevant quality standards throughout the supply chain. It involves establishing quality standards, monitoring and controlling processes, and implementing corrective actions to maintain and improve product quality.

» **Supplier selection and qualification:** Quality assurance begins with selecting reliable and reputable suppliers. Organizations need to assess potential suppliers based on their quality management systems, certifications, track records, and adherence to industry standards

» **Quality standards and specifications:** Clear quality standards and specifications should be defined for each product or service in the supply chain. This includes setting criteria for product attributes, performance, functionality, and any relevant regulatory requirements

» **Incoming inspection:** Incoming inspection involves checking and verifying the quality of raw materials, components, and parts received from suppliers. This ensures that the incoming materials meet the specified quality standards before they are used in production

» **Process control:** Quality assurance includes implementing effective process controls throughout the manufacturing or service delivery processes. This may involve using statistical process control techniques, monitoring key performance indicators, and conducting regular inspections or audits to identify and address any deviations or non-conformities

» **Continuous improvement:** Quality assurance is an ongoing process of continuous improvement. Organizations should regularly review their quality management systems, collect and analyze data, and identify areas for improvement. This may involve implementing corrective actions, conducting root cause analysis, and optimizing processes to enhance quality and efficiency

» **Traceability and documentation:** Maintaining traceability and documentation is crucial for quality assurance. It involves tracking and documenting the movement of products, components, and materials throughout the supply chain. This allows for effective identification, investigation, and recall of any defective or non-compliant products

» **Training and awareness:** Quality assurance requires a well-trained and knowledgeable workforce. Providing appropriate training and awareness programs on quality standards, procedures, and best practices is essential to ensure that employees understand their roles and responsibilities in maintaining and improving product quality

» **Collaboration with suppliers:** Collaboration and communication with suppliers are important for quality assurance. Establishing clear expectations, sharing information, and maintaining open lines of communication help to address quality issues proactively and foster a culture of quality throughout the supply chain

» **Customer feedback and satisfaction:** Quality assurance should incorporate mechanisms for gathering and analyzing customer feedback. This information helps identify areas of improvement and guides decision-making processes to enhance customer satisfaction and loyalty

Quality Assurance and Consumer Expectations

Quality assurance and consumer expectations are closely linked in the world of products and services. Meeting consumer expectations is crucial for the success and reputation of a business. Consumers today have become more discerning and have higher expectations when it comes to the quality of products and services they purchase. They expect products to be reliable, durable, and perform as advertised. They also expect services to be delivered in a timely and efficient manner, with excellent customer support.

Quality assurance plays a vital role in meeting consumer expectations. By implementing robust quality control processes throughout the product or service lifecycle, companies can ensure that their offerings meet the desired level of quality. To effectively meet consumer expectations, companies need to understand their target market and gather insights into customer preferences and demands. This can be achieved through market research, customer surveys, feedback mechanisms, and social listening. By understanding consumer expectations, companies can align their quality assurance efforts to deliver products and services that match or exceed those expectations. Moreover, consumer expectations are not static and can evolve over time. Technological advancements, industry trends, and changing customer needs can all influence consumer expectations. Therefore, businesses must stay agile and adaptable, continuously improving their quality assurance processes and keeping up with evolving consumer demands. By staying attuned to consumer expectations and proactively addressing them, companies can build customer loyalty, gain a competitive edge, and foster long-term success.

Global Regulatory Frameworks and Standards

Regional and national regulatory frameworks play a crucial role in shaping the operations and practices of agribusinesses. These frameworks are designed to ensure food safety, protect the environment, promote fair trade, and safeguard the interests of consumers and farmers.

1. **FAO's work on food safety and quality:** Unsafe food poses a significant threat to human health and global economies, with an estimated 600 million cases of foodborne illnesses occurring each year. Consequently, prioritizing food safety becomes a crucial public health imperative and an essential component in attaining food security.

   FAO is globally acknowledged for its leadership in formulating international food safety initiatives and effectively implementing them at the national level. The Food Safety and Quality Programme promotes a comprehensive and interdisciplinary approach to managing food safety and offers practical and viable "food chain" solutions to address specific food safety challenges, aligning with FAO's Strategy for Enhancing Global Food Safety. In keeping with this year's World Food Safety Day theme, the piece elaborates on how sound science and strong collaboration drive the development of food standards, the challenges around their implementation and the need to keep standards relevant by staying at par with scientific advances. Achieving food security for all is at the heart of FAO's efforts to make sure people have regular access to enough high-quality food to lead active, healthy lives (www.fao.org).

The Food Safety and Quality Unit focuses on strengthening food safety and quality control systems at national, regional, and international levels. This involves enhancing regulatory capacities, supporting capacity development for food control and safety management, providing scientific advice for food safety standards, promoting risk-based food safety management, facilitating communication and access to information, collecting and analyzing food chain intelligence, and evaluating new technologies for improved food safety and public health protection.

2. **World Health Organization (WHO):** WHO's role in food safety is to reduce the burden of food-borne disease by advising and assisting member countries to reduce exposure to unacceptable levels of microorganisms or chemicals in food. Its work includes international standard setting. WHO also provides technical assistance to governments to ensure supply of safe food to their people. Many WHO activities in food safety and quality are carried out in close collaboration with FAO. These agencies have developed several technical papers on aspects related to food safety and quality assurance. Further, FAO and WHO jointly undertake to conduct risk assessment activities through their permanent Committees/Groups, namely the Joint FAO/WHO Expert Committee on Food Additives and Contamination (JECFA), the Joint FAO/WHO Meeting on Pesticide Residues (JMPR) and the Joint FAO/WHO Meeting on Microbiological Risk Assessment (JEMRA), and other expert consultations.

3. **World Trade Organization (WTO):** With respect to food safety and quality matters, the Sanitary and Phytosanitary Measures (SPS) and Technical Barriers to Trade (TBT) agreements are important. The SPS

agreement covers health-related aspects of food, such as food additives, levels of contaminants, codes of hygienic practices and packaging materials in contact with food. These should be based on risk assessment. The TBT agreement covers technical standards of food, such as product description, labelling, nutrition, method of analysis and package specifications.

4. **Codex Alimentarius Commission (CAC):** U. S. international food standards, guidelines and codes of practice contribute to the safety, quality and fairness of this international food trade. Consumers can trust the safety and quality of the food products they buy and importers can trust that the food they ordered will be in accordance with their specifications. The Codex Alimentarius is a collection of internationally adopted food standards and related texts presented in a uniform manner. These food standards and related texts aim at protecting consumers' health and ensuring fair practices in the food trade. The publication of the Codex Alimentarius is intended to guide and promote the elaboration and establishment of definitions and requirements for foods to assist in their harmonization and in doing so to facilitate international trade. The primary objectives of the programme are to protect the health of consumers and ensure fair practice in food trade.

» **Protecting Consumer Health:** Public concerns about food safety issues often place Codex at the center of global debates. Veterinary drugs, pesticides, food additives and contaminants are some of the issues discussed in Codex meetings. Codex standards are based on sound science provided by independent international risk assessment bodies or ad-hoc consultations organized by FAO and WHO.

» **Removing Barriers to Trade:** The reference made to Codex food safety standards in the WTO's Agreement on Sanitary and Phytosanitary measures (SPS Agreement) means that Codex has far reaching implications for resolving trade disputes.

5. **Quality Management Systems (QMS):** A Quality Management System (QMS) is a collection of business processes aimed at consistently meeting customer requirements and enhancing satisfaction. It encompasses organizational goals, policies, processes, documented information, and resources needed for implementation and maintenance. Originally focused on predictable outcomes and statistical sampling, modern QMS emphasizes team cooperation, continual improvement, and integration with sustainability and transparency initiatives. The ISO 9000 family of standards is widely

implemented worldwide, and other QMS frameworks like Natural Step prioritize sustainability and systematic thinking to address quality issues.

QMS, emphasizing its role in ensuring quality assurance and continuous improvement throughout the agribusiness operations. ISO 9001:2015 is the most recognized and implemented quality management system standard in the world. Other standards related to quality management systems include the rest of the ISO 9000 series (including ISO 9000 and ISO 9004), the ISO 14000 series (environmental management systems), ISO 13485 (quality management systems for medical devices), ISO 19011 (auditing management systems), and IATF 16949 (quality management systems for automotive-related products). Implementation of QMS in agribusiness such as consistent product quality, improved customer satisfaction, and efficient resource management, resistance to change, resource constraints, and maintaining the effectiveness of the system.

6. **Food Safety Management Systems (FSMS):** Food Safety Management System (FSMS) is a set of policies, procedures, and practices implemented by food businesses to ensure the safety and integrity of food products throughout the entire supply chain. The primary goal of an FSMS is to identify and control food safety hazards, prevent foodborne illnesses, and comply with applicable regulations and standards. Key components of an FSMS may include prerequisite programs (PRPs), such as Good Manufacturing Practices (GMP), sanitation, and hygiene practices, as well as Hazard Analysis Critical Control Point (HACCP) principles. Implementing an FSMS enables food businesses to systematically manage food safety risks, demonstrate their commitment to consumer protection, and build trust among customers, regulators, and stakeholders.

Prerequisite programmes (PRPs) are defined by the ISO as basic conditions and activities which are necessary in an organization and throughout the food chain to maintain food safety. According to the Food and Agriculture Organization, examples of prerequisite programmes are:

» **Good Agricultural Practices (GAP):** According to FAO, GAP is a "collection of principles to apply for on-farm production and postproduction processes, resulting in safe and healthy food and non-food agricultural products, while taking into account economic, social and environmental sustainability." Implementing Good Agricultural Practices (GAP) during on-farm production and post-production processes resulting in safe agricultural products is of immense importance for ensuring a safe food supply. Many importing countries as well as domestic buyers, especially organized retailers, are requiring producers to implement GAP as a prerequisite for procurement to ensure the quality and safety of their produce. In addition, implementing GAP also helps promote sustainable agriculture and contributes to meeting national and international environmental and social developmental objectives. It has been documented that implementation of GAP encourages promotion of the optimum use of resources such as pesticides, fertilizers, and water, and eco-friendly agriculture. Its social dimension would be to protect the agricultural workers' health from improper use of chemicals and pesticides (FAO,2016).

Good Agricultural Practices (GAP) are of paramount importance in ensuring food safety and quality throughout the agricultural production process. By emphasizing practices that prevent contamination, GAP reduces the risk of microbial and chemical hazards. GAP also encourages sustainable agricultural practices, protects worker health and safety, and meets consumer expectations for safe and high-quality food. Adhering to GAP principles allows agribusinesses to establish a comprehensive framework that safeguards the integrity of their products, contributing to a resilient and sustainable food system.

Implementing GAP in agribusiness brings benefits like improved product quality, market access, and consumer confidence. However, challenges include resource constraints and the need for training. GAP offers enhanced food safety, market opportunities, trust, and sustainability, but initial costs, training, regulatory changes, and supply chain consistency are hurdles to overcome.

» **Good Manufacturing Practices (GMP):** Good Manufacturing Practices (GMP), also known as cGMP (current Good Manufacturing Practice), is a quality assurance framework that ensures consistent production and control of medicinal products according to their intended use and product specifications. GMP sets quality standards for production and quality control, encompassing defined processes, validation, documentation, suitable facilities, personnel, and materials. It includes legal aspects such as responsibilities for distribution, contract manufacturing, testing, and addressing product defects and complaints (Anonymous, 2018).

GMP ensures food safety and quality in processed food products by setting rigorous standards for production. It prevents contamination, maintains consistency, and meets regulatory requirements. Implementing GMP in agribusiness delivers safe products, meets customer expectations, and ensures regulatory compliance. Challenges include training, resource allocation, and ongoing compliance.

» **Hazard Analysis and Critical Control Points (HACCP):** HACCP (Hazard Analysis and Critical Control Points) is a proactive food safety system in which each stage of food production, as well as the storage and distribution of food products, undergoes scientific examination and regulation to address biological, chemical, and physical risks. This system ensures that the products are safe and of good quality. The system is extremely desirable in view of the changing scenario in the international trade. The Ministry provides grant of 50 per cent subject to a limit of Rs.10 lakhs towards the cost of implementing Total Quality Management (TQM) including HACCP and ISO-9000 certifications (Ibrahim, 2020).

There are seven basic HACCP principles which are developed and implemented internationally are as follows;

i. Identify hazard analysis accessing less risk
ii. Identifying Critical Control Points (CCPs)
iii. Establishing Critical Limits (CL)
iv. Implementing continuous Monitoring
v. Implementing Corrective Measures
vi. Verification and Validation
vii. Documentations and Records keeping

HACCP implementation process outlines the step-by-step process of implementing HACCP, starting from assembling a HACCP team to conducting a hazard analysis, establishing critical control points, and implementing effective monitoring, corrective actions, and verification procedures. Implementing HACCP in agribusiness operations are explored, such as improved food safety, enhanced product quality, and increased consumer confidence. And some of the challenges includes data collection and analysis, resource allocation, and maintaining HACCP compliance.

Emerging Trends in Quality Assurance and Food Safety

» **Blockchain Technology:** Blockchain is emerging as a promising technology for enhancing transparency and traceability in the food supply chain. It enables secure, decentralized record-keeping, allowing stakeholders to track and verify the origin, handling, and quality of food products

- **Advanced Testing and Detection Methods:** Rapid advancements in testing and detection methods offer improved capabilities for identifying foodborne pathogens, contaminants, and allergens. Technologies such as DNA-based testing, spectroscopy, and biosensors provide faster, more accurate, and cost-effective analysis
- **Data Analytics and AI:** The use of data analytics and artificial intelligence (AI) can revolutionize quality assurance and food safety. AI algorithms can analyze large volumes of data to identify patterns, predict risks, and optimize processes. This enables proactive risk management and more efficient resource allocation
- **Internet of Things (IoT):** IoT devices and sensors can monitor various parameters, such as temperature, humidity, and storage conditions, in real-time throughout the supply chain. This enables early detection of deviations, timely interventions, and improved overall control over product quality and safety
- **Enhanced Food Safety Culture:** Developing a strong food safety culture within organizations is gaining attention. It involves fostering a shared commitment to food safety among employees, promoting training and awareness programs, and establishing robust systems for reporting and addressing safety concerns
- **Sustainability and Environmental Impact:** There is a growing emphasis on sustainable practices and reducing the environmental impact of the food industry. Quality assurance and food safety strategies are increasingly incorporating sustainable sourcing, eco-friendly packaging, and waste reduction measures
- **Novel Food Technologies:** Advances in food technologies, such as high-pressure processing, pulsed electric field technology, and cold plasma treatment, offer alternative methods for enhancing food safety and extending shelf life without the use of traditional preservatives.

Conclusion and Recommendations

Food safety and nutrition are intricately linked, as unsafe food initiates a detrimental cycle of disease and malnutrition that adversely affects vulnerable groups such as infants, young children, the elderly, and the sick. Given that food supply chains span multiple national and regional boundaries, collaborative efforts among governments, producers, suppliers, distributors, and consumers are imperative to ensure food safety in the modern era. Food safety, a scientific field, encompasses the proper handling, preparation, and storage of food to prevent foodborne illnesses. The growing global concern regarding food safety and quality, particularly concerning human health, has prompted numerous countries to implement quality control programs encompassing all food ingredients, including those from animal sources. This necessitates a series of routine activities that must be followed to avert the potential occurrence of severe health risks.

Quality assurance and food safety are essential components of agribusiness operations. By prioritizing these aspects, companies can protect consumer health, build trust, comply with regulations, prevent recalls, enhance product quality, and facilitate market access. These measures not only benefit the agribusinesses themselves but also contribute to the overall safety and well-being of consumers worldwide. Maintaining high-quality standards in the agricultural industry is crucial for consumer health and safety, reputation building, market access, environmental sustainability, economic impact, and regulatory compliance. It is a comprehensive approach that benefits farmers, consumers, the industry, and the overall well-being of society. They ensure consumer safety, regulatory compliance, brand reputation, risk mitigation, supply chain management, and facilitate international trade. By prioritizing these aspects, agribusinesses can enhance their competitiveness, protect consumers, and contribute to a sustainable and responsible food industry.

Quality assurance and consumer expectations are interconnected aspects of delivering high-quality products and services. Meeting and exceeding consumer expectations requires companies to implement robust quality assurance processes, understand customer preferences, and continuously improve to stay ahead in a dynamic market. By doing so, businesses can build trust, satisfy customers, and differentiate themselves from competitors.

There is no doubt that food and agribusiness firms have increasingly to deal with competitive markets in which market directedness of produce and services and the efficiency and reliability of their delivery become decisive aspects for competitiveness. With regard to these developments, firms are implementing quality assurance systems, such as Good Manufacturing Practice (GMP), hazard analysis critical control point (HACCP), International Standard Organization 9000 (ISO) and Total Quality Management (TQM) Systems. By implementing these quality assurance systems, food and agribusiness firms can enhance their reputation, meet regulatory requirements, improve operational efficiency, and differentiate themselves in the competitive market, ultimately contributing to their overall competitiveness and success.

References

Addis, M., & Sisay, D., (2015). A Review on Major Food Borne Bacterial Illnesses. *J. Trop. Dis.*, **3**: 176.

Anonymous, (2018). Report on WHO Workshop on Implementation of Good Manufacturing Practices (GMP) for Biological Products, Seoul, Republic of Korea.

Battu, R., Singh, S. B., & Kang, B. K., (2004). Contamination of liquid milk and butter with pesticide residues in the Ludhiana district of Punjab state, India. *Ecotoxicol. Environ. Saf.*, **59**: 324-331.

Bergdoll, M. S., & Lee, W. A. C., (2006). Staphylococcal intoxications. *Foodborne Infect Intox.*, **3**: 523-552.

CFSAN (Center for Food Safety and Applied Nutrition), (2007). Hazards & Controls Guide for Dairy Foods. HACCP Guidance for Processors Version 1.1 June 16, 2006. Center for Food Safety and Applied Nutrition, USA.

EFSA (European Food Safety Authority), (2005). Statement of the Scientific Panel on Contaminants in the Food Chain to a summary report on Acrylamide in food of the 64th meeting of the joint FAO/WHO Expert Committee on food additives. The EFSA J., 619, 12.

Food and Agriculture Organization, (1997). Gender: the key to sustainability and food security. SD Dimensions, May 1997. Rome, Sustainable Development Department.

Food and Agriculture Organization, (2003). Workshop on the prospective on the application of lactoperixidase system in milk handling and preservation in Indonesia. FAO. Jakarta.

Fung, F., Wang, H. S., & Menon, S., (2018). Food safety in the 21st century. *Biomedical Journal*, **41**: 88-95.

Gemeda, F., (2023). Review on Food Safety, Quality Assurance and Control. *IAR Jr. Agr Sci Fd Res*, **3** (2): 1-9.

Ibrahim, O. O., (2020). Introduction to Hazard Analysis and Critical Control Points (HACCP). *EC Microbiology*, **16** (3): 1-7.

www.faostat.com
www.fao.org
www.indiadairy.com

14

Rural Development and Agribusiness Integration

Dr. Yash Unjia[1*], Dr. Savan Padaliya[2], Yogi Agrawat[3], Meera Padaliya[4]

[1*]Project Head-FPO, Indiagro Consortium Producer Company Limited
[2]Assistant Professor, Department of Agribusiness Management, Ganpat University
[3]Assistant Professor, Department of Pharmaceutical Management, Ganpat University
[4]Ph.D Scholar, Dept. Of Agricultural Economics, NAU, Gujarat
*Corresponding author Mail Id: yashunjia003@gmail.com

Abstract

Significant portion of India's population residing in rural areas and depending on agriculture for their sustenance. Prosperity of rural India are closely intertwined with the success of the agricultural sector. It aims to enhance the quality of life for rural communities by addressing various challenges they face, such as limited access to basic services, inadequate infrastructure, poverty, unemployment, and lack of educational opportunities. Integrating agribusiness into rural development initiatives exploring how it can promote employment, stimulate entrepreneurship, and drive economic growth in rural areas. This chapter discussed Economic Diversification, Value Chain Development, Market Access and Demand Orientation, Skill Development and Capacity Building, Access to Finance and Support Services and Policy and Institutional Support for rural development through Agribusiness. It also discussed emerging approach like Agro-Tourism, Commercialization of Agriculture and Encouraging Agro-Processing and Value Addition.

Keywords: Agribusiness, Rural development, Sustainable development, Agro tourism

Introduction

In developing nations, more than half of total populations are engaged in agriculture and many of them living in abject poverty. These factors imply that agricultural and rural development must work together to assist developing nations in addressing the major issues of food security, extreme poverty, and slow economic growth. Rural development and agribusiness sector integration helps to uplift socio-economic infrastructures in rural areas with focusing on agriculture development. In India, where most people live in rural areas and depend on agriculture for their living, agriculture is essential for rural development. Due to a lack of commercialization and subsistence farming, rural areas in India endure extreme poverty. India needs to work on the inclusive commercialization of agriculture to succeed in rural development. This chapter will explore several aspects of agribusiness and rural development integration in India and suggest a comprehensive approach for achieving long-term development in rural areas. The government keeps enacting new policies and initiatives to help farmers, boost productivity and ensure the country's food security.

Agriculture: The Backbone of India's Rural Economy

For many years, agriculture has been the foundation of rural India's economy, serving as a key factor in generating employment opportunities, maintaining food security, and fostering growth in the economy. The growth and prosperity of rural India are directly linked to the performance of the agricultural sector because a significant portion of India's population lives in rural areas and relies on agriculture for their subsistence.

Historical Significance

India's history and culture are deeply rooted in agriculture and go back thousands of years. The Indian subcontinent is frequently referred to as the "Cradle of Agriculture," as it was among the first places where people started participating in organised farming. In the Indus Valley Civilization (3300 BC–1300 BC), Vedic Period (1500 BC–600 BC), Mauryan Empire (322 BC–185 BC), Gupta Empire (320 CE–550 CE), and British Colonial Rule (1757–1947), agriculture may be found as a key activity. The rural way of life is shaped by the traditional farming methods and agricultural knowledge that have been passed down through the years. Agriculture has seen remarkable changes, becoming more productive and resilient, from the Green Revolution in the 1960s to modern technology breakthroughs.

Livelihoods and Employment

A large section of rural India's population depends on agriculture for both direct and indirect means of subsistence. Farmers and their families are involved in a variety of agricultural pursuits, including as farming, raising livestock, and fishing. The agricultural value chain also creates employment opportunities across related industries including processing, marketing, transportation, and agribusiness. agriculture's involvement in rural employment is essential. agriculture's involvement in rural employment is essential for social inclusion, poverty reduction, and maintaining a healthy balance between rural and urban areas.

Food Security

A large section of rural India's population depends on agriculture for both direct and indirect means of subsistence. Farmers and their families are involved in a variety of agricultural pursuits, including as farming, raising livestock, and fishing. The agricultural value chain also creates employment opportunities across related industries including processing, marketing, transportation, and agribusiness. agriculture's involvement in rural employment is essential. agriculture's involvement in rural employment is essential for social inclusion, poverty reduction, and maintaining a healthy balance between rural and urban areas.

Economic Contribution

India's rural economy and GDP as a whole both benefit greatly from agriculture. It provides money, increases purchasing power for rural residents, and encourages economic activity there. The agriculture industry contributes to value addition and the creation of jobs by providing raw materials to sectors including textile, food, and agro-based manufacturing. Additionally, agricultural exports boost the nation's economic resilience by adding to foreign exchange revenues.

Challenges and Resilience

India's agricultural sector has a number of challenges, such as scattered landholdings, water scarcity, effects of climate change, market fluctuations, and limited access to capital and modern innovations. Despite these difficulties, Indian farmers have shown great fortitude in adjusting to shifting conditions, implementing novel practises, and diversifying their sources of revenue. In order to increase the resilience of Indian agriculture, government policies, research and development, and institutional assistance have all been extremely important.

Rural Development

The process of enhancing rural areas' economic, social, and environmental situations is referred to as rural development. It seeks to improve the condition of life for rural populations by tackling a number of issues they encounter, including restricted access to essential services, insufficient infrastructure facilities, low incomes, and a lack of opportunities for education.

Some important aspects of rural development are as follows:

1. Infrastructure Development: Infrastructure development in rural areas is essential for their overall growth. This entails developing and enhancing roads, bridges, water delivery systems, hospitals, schools, and other essential facilities.

2. Agriculture and Livelihood Development: Promoting agricultural growth is crucial since it is frequently the main source of income for rural areas. It includes Implementing advanced farming methods, granting access to finance and information about markets, setting up irrigation systems, and promoting agricultural diversification.

3. Employment and Skill Development: Creating jobs in rural areas helps with lowering poverty and decreasing migration to metropolitan areas. Rural economic development can be aided by entrepreneurship promotion, career training programmes, and assistance for small and medium-sized businesses (SMEs).

4. Access to Basic Services: Providing rural populations with access to essential services like medical care, schooling, safe drinking water, and sanitation is essential for raising their standard of living. Governments and development agencies seek to construct schools, hospitals, and infrastructure for safe drinking water and sanitation.

5. Technology and Digital Connectivity: By bridging the digital gap and giving remote places access to the internet and other forms of digital connectivity, we may create new possibilities for e-learning, e-health, e-commerce, and knowledge sharing. Additionally, it can help rural areas observe a rise in digital entrepreneurship.

6. Natural Resource Management: Sustainable and environmentally responsible rural development should be practised. In addition to improving livelihoods, encouraging practises like sustainable farming, watershed management, forest protection, and renewable energy can help to protect natural resources.

7. Governance and Institutional Development: Sustainable rural development depends on bolstering local government systems, empowering rural populations, and including them in decision-making procedures. This may entail fostering social inclusion, developing participatory methods, and strengthening capability.

Government initiatives, development projects, public-private partnerships (PPP), and community engagement are frequently used to carry out rural development programmes. Depending on the individual requirements, priorities, and features of each rural region, several techniques and approaches may be used.

Significance of Agribusiness Integration for Rural Development

1. Supporting Employment and Ancillary Businesses

The incorporation of agriculture is crucial for providing employment possibilities and growing ancillary businesses in the context of rural development. Strategically merging agricultural cultivation, processing, marketing, and distribution activities into rural development strategies is known as agribusiness integration. This chapter focuses on the importance of incorporating agriculture into rural development projects and examines how it may foster entrepreneurship, increase employment, and promote economic development in rural regions. Government initiatives, construction initiatives, public-private partnerships (PPP), and community engagement. Depending on the individual needs, priorities, and features of each rural region, several techniques and approaches may be used.

2. Economic Diversification

Rural areas can diversify their economies by integrating their agribusinesses. Rural communities can benefit from value-added enterprises like food processing, agro-based manufacturing, and agro-tourism by pursuing activities other than traditional agriculture. These businesses offer rural communities both skilled and unskilled labour employment options. Economic diversification strengthens rural economies by reducing reliance on just agriculture.

3. Value Chain Development

The entire agricultural supply chain, from production to consumption, must be improved and strengthened as part of agribusiness integration. This entails boosting post-harvest management, investing in processing facilities and creating effective marketing and distribution channels, among other improvements to agricultural practises. employment can be created at many phases of the value chain, including farm management, industrial operations, transportation, and retail management. It promotes entrepreneurs and aids in the expansion of ancillary companies which support the agribusiness sector.

4. Market Access and Demand Orientation

Market access and a production strategy focused on consumer demand are supported by agribusiness integration. Rural farmers and business owners can adapt to individual client demands and preferences by matching production to market demand. Because there is a greater demand for high-quality produce as a result of the transition towards market-oriented agriculture, there is an increase in the demand for skilled labour in the production, processing, packaging, and marketing sectors. It promotes creativity and competition, which results in sustainable job creation.

5. Skill Development and Capacity Building

Focusing on skill and capacity development is necessary for integrating agribusiness into rural development. Farmers, agribusiness owners, and rural entrepreneurs can improve their knowledge and competencies by receiving training and technical support. Programmes for skill development can cover topics including contemporary farming methods, value addition, quality assurance, business management, marketing, and entrepreneurship. Rural residents are better able to participate in agriculture operations and take advantage of employment opportunities when they are equipped with the necessary skills.

6. Access to Finance and Support Services

Access to funding and support services are necessary for agribusiness integration to support employment and ancillary enterprises. Initiatives for rural development should make it easier for agriculture firms to obtain credit, microfinance, and investment opportunities. Additionally, offering mentoring programmes, market information, and assistance with business growth can help entrepreneurs build strong business plans, improve their operations, and broaden their market. The expansion and viability of rural businesses is made possible by access to financing and support services, increasing employment opportunities.

7. Policy and Institutional Support

Effective agribusiness integration for rural development requires favourable policies and institutional support. Governments and development organisations should create policies that encourage the expansion of agribusiness, promote a positive economic climate, and remove regulatory barriers. This includes establishing specialised institutions, providing technical assistance, and encouraging public-private partnerships in order to support agribusiness activities. Policymakers should place a high focus on building out rural infrastructure, such as storage facilities, transportation systems, and market links, to assist agricultural activity.

Commercialization of Agriculture: A Key to Rural Development

Agriculture's commercialization can in fact be a major force behind rural development. Traditional subsistence farming can be converted into commercial agriculture, which has a number of advantages for rural communities that support their overall growth and prosperity. The following are a few ways that commercialising agriculture can promote rural development:

Increased income and employment opportunities: Commercial agriculture involves raising plants, animals, and other products for market sales, which can greatly boost farmers' profits. As a result, rural people have a source of income, which helps to reduce rural poverty. More employment opportunities arise as agricultural production increases, both directly in farming activities and indirectly in adjacent businesses like processing, storage, transportation, and marketing.

Technological advancements and knowledge transfer: Modern farming methods, tools, and best practises are frequently needed in commercial agriculture. As a result, productivity, profitability, and efficiency all increase. Farmers are more likely to use training, information, and support services offered by agricultural extension agencies, research institutes, and private businesses as they engage in commercial operations. This knowledge transfer improves farmers' abilities, advances farming methods, and encourages innovation in rural areas.

Infrastructure development: Rural infrastructure, such as roads, irrigation systems, storage facilities, and processing units, must be developed for commercial agriculture. In addition to assisting agricultural activity, these infrastructure expenditures also benefit the larger rural population. Enhancing market access, lowering post-harvest losses, facilitating commodities movement, and luring other industries and services to rural regions are all benefits of improved infrastructure.

Market integration and value addition: Farmers can connect with larger markets outside of their immediate area thanks to commercialization. It promotes the creation of effective marketing channels, such as supply chains, agribusinesses, and cooperatives. Farmers that have access to larger markets and can sell their produce for more money and benefit from a steadier market for their produce. Additionally, commercial agriculture promotes value addition through packaging and processing, which raises farmers' incomes and opens doors for rural agro-industries.

Rural entrepreneurship and agribusiness development: The expansion of agribusiness firms is prompted by the commercialization of agriculture. Farmers may diversify their businesses and take part in ventures like agro-processing, farm input distribution, farm equipment rental, and agritourism as farming becomes more market-oriented. By creating more jobs, income, and tax revenues, these entrepreneurial endeavours support the growth of rural economies.

Improved social services and quality of life: Commercial agriculture's economic prosperity can be invested back into rural areas, enhancing infrastructure and social services. The funding for education, healthcare, sanitization, power, and clean water supply is up to local governments. This improves rural residents' quality of life and lessens the gap between rural and urban communities.

It's crucial to remember that agriculture should be commercialised in a way that is both inclusive and sustainable. The advantages should be distributed fairly, the environment should be protected, responsible land usage should be encouraged, and small-scale farmers' needs and goals should be taken into account. Furthermore, to help farmers effectively transition to commercial agriculture and fully realise its potential for rural development, supportive policies, access to finance, and capacity-building programmes are essential.

Encouraging Agro-Processing and Value Addition

» Integrating agribusiness with agro-processing and value-added enterprises has become a potent tactic in an effort for rural development. Rural communities can open up new opportunities for economic growth, the creation of jobs, and general development by encouraging farmers to adopt value-added industries in place of conventional farming practises. The transforming potential of agro-processing and value addition in promoting rural development and supporting agribusiness integration.

» **Economic Growth and Income Generation:** Rural communities have prospects for economic growth due to agroprocessing and value addition businesses. Farmers can gain access to more valuable markets and command higher prices for their crops by adding value through processing. Farmers benefit from higher incomes and better financial health as a result of this.

The development of processing facilities and value-added businesses also encourages rural entrepreneurship, creating job opportunities and supporting a healthy business ecosystem.

» **Market Expansion and Access**: Expanding their market reach is made possible for rural farmers through agroprocessing and value addition. Agricultural products that have been processed have a longer shelf life, making them ideal for shipping to far-off markets. Farmers can meet changing consumer expectations and capitalise on specialised markets by expanding their product line and providing commodities with high value-added features. Agro processing also makes it easier to build connections and linkages with other market participants, giving farmers direct access to retailers, exporters, and other market participants and bolstering their competitiveness.

» **Reduction of Post-Harvest Losses**: Agroprocessing is essential for reducing post-harvest losses, a problem that persists in rural agriculture. Farmers can increase the shelf life and maintain the quality of their goods by converting perishable agricultural commodities into value-added products. This increases the use of agricultural resources while minimising losses and ensuring food security. Farmers may improve the shelf life and marketability of their crops by efficient processing processes like drying, canning, and freezing, thus boosting their profitability.

» **Product Diversification and Consumer Demand**: Farmers are able to expand their product offerings and meet a wide range of consumer preferences by adding value through agroprocessing. Farmers may satisfy the changing needs of consumers for ease of access, variety, and nutritional value by processing raw materials into processed foods like juices, jams, sauces, and snacks. Additionally, value-added goods can be tailored to target particular consumer groups, such as those who prefer organic, gluten-free, or ethically sourced items. This diversification boosts market competition and opens doors for niche marketing and market segmentation.

» **Technology Adoption and Innovation**: Modern technologies and creative methods must be adopted for agroprocessing and value addition. Farmers are encouraged to adopt technical innovations in packaging methods, quality control procedures, and processing equipment through this integration. Adopting effective processing techniques boosts efficiency while also guaranteeing product safety and quality standards. Agroprocessing innovation can also result in the creation of new product lines, the use of byproducts, and the investigation of market trends.

» **Capacity Building and Knowledge Transfer**: successful Initiatives for agroprocessing and value addition that are rely on knowledge transfer and capacity building in rural areas. Farmers are equipped with the requisite skills in processing methods, quality control, marketing, and company management through training programmes, workshops, and knowledge-sharing platforms. Collaborations among farmers, researchers, and industry professionals also help to share knowledge and best practises, enabling farmers to use productive and sustainable processing techniques.

Agro-Tourism: A Sustainable Option for Rural Development

» Agrotourism is now understood to be a potent instrument for fostering rural development and incorporating agribusiness into regional economies. Agrotourism, which combines agriculture, tourism, and entrepreneurship, provides rural communities with a sustainable approach to make the most of their agricultural resources, make money, and promote economic progress.

» **Agro-Tourism as a Rural Development Strategy**: Agro-tourism is characterised as a kind of tourism that blends rural experiences, agricultural pursuits, and cultural immersion. Visitors get the chance to actively participate in agricultural activities and discover the rural way of life. The section outlines the potential advantages of agrotourism, including the diversification of the rural economy, the creation of new revenue sources, and the promotion of infrastructural growth.

» **Agribusiness Integration in Agro-Tourism**: Agro-tourism initiatives that incorporate agribusiness strengthen value chains, increase market access for agricultural products, and give farmers more opportunity to make money. It emphasises the value of working together and forming alliances with government organisations, agribusiness players, and travel operators.

» **Agro-Tourism: A Driver for Rural Entrepreneurship:** Agrotourism provides farm-based lodging, guided tours, artisanal product creation, and cultural experiences to enable rural residents to start their own businesses. It emphasises on how crucial it is to provide rural entrepreneurs with the marketing, customer service, and sustainable farming practises expertise they need through capacity building programmes, workshops, and knowledge-sharing platforms. It also emphasises how agrotourism has the ability to empower rural youth and women by promoting intergenerational knowledge transfer, gender equality, and inclusive development.

» **Sustainable Agro-Tourism Practices:** In order to be sustainable over the long term, agrotourism must place a high priority on environmental protection, cultural preservation, and community involvement. It places a focus on the necessity of sustainable tourism practises, tourist education, and awareness campaigns to encourage adherence to regional traditions, ecological awareness, and responsible travel practises. The part also looks at how agrotourism encourages community involvement by including regional communities in the planning and implementation of agrotourism initiatives.

Others

» **Skill Development and Capacity Building:** The sustainable growth of India's rural sector depends on the development of skills and capacities. Farmers and rural communities can be empowered to seize new opportunities in agriculture and allied businesses by receiving training and instruction in modern farming methods, agribusiness management, and entrepreneurship. To guarantee that farmers and rural youth have access to the skills and knowledge they need for their growth, it is crucial to construct vocational training facilities and agricultural extension services in rural regions.

» **Access to Finance and Credit Facilities:** Another important element influencing rural development in India is access to credit and financial resources. By ensuring that farmers and rural business owners have inexpensive credit and financial services available to them, we can help them invest in their enterprises and strengthen their financial position. Policymakers should concentrate on fostering financial inclusion in rural areas and providing a supportive climate for rural financial institutions.

» **Research and Development in Agriculture:** Investing in research and development (R&D) in agriculture is crucial for enhancing productivity, resilience, and sustainability in India's rural sector. R&D efforts should focus on developing climate-resilient crop varieties, promoting sustainable farming practices, and improving post-harvest management and storage facilities. Establishing research institutions and promoting collaboration between researchers, farmers, and the private sector can help accelerate the adoption of innovative agricultural technologies and practices in rural areas.

» **Women Empowerment and Gender Equality:**For India's rural development to be inclusive and sustainable, it is crucial to promote gender equality and women's empowerment in rural regions. The empowerment of women can improve food security, nutrition, and the general well-being of rural communities. Women play a vital role in agriculture and rural economies. Enhancing women's access to education, skills, supplies, and decision-making possibilities in the rural sector should be the main goal of policies and programmes.

Conclusion

Rural development and agribusiness integration in India require a comprehensive strategy to effectively handle the commercialization of agriculture, infrastructure development, skill development, access to finance, research and development, and women's empowerment, . India can make sustainable growth in rural areas and improve the lives of millions of people who depend on agriculture by implementing a comprehensive strategy and encouraging cooperation among stakeholders.

References

Bairwa, S. L., Kalia, A., Meena, L. K., Lakra, K., & Kushwaha, S. (2014). Agribusiness management education: a review on employment opportunities. *International Journal of Scientific and Research Publications (IJSRP)*, *4*(2), 1-4.

Bhadauria, A., & Rastogi, H. (2012). Rural Tourism: An Avenue for Sustainable Rural Development. *International Journal of Multidisciplinary Research*, *1*(8), 2277-9302.

Chandra, D. G., & Malaya, D. B. (2011, April). Role of e-agriculture in rural development in Indian context. In *2011 International Conference on Emerging Trends in Networks and Computer Communications (ETNCC)* (pp. 320-323). IEEE.

Davis, J., & Tschirley, D. (Eds.). (2017). Agricultural value chain finance: Tools and lessons. *Food and Agriculture Organization of the United Nations*. Retrieved from http://www.fao.org/3/a-i7777e.pdf

Dixon, C. (2015). *Rural development in the third world*. Routledge.

Gupta, V. (2015). Retail Operating Models in Indian Agri-business Sector. *Journal of Rural Development*, *34*(1), 49-60.

Hans, V. (2008). Agri-Business and Rural Management in India-Issues and Challenges. *Agri-Business and Rural Management in India-Issues and Challenges (September 21, 2008).*

Hans, V. B. (2006). Agribusiness in India: Issues and challenges. *Journal of Global Economy*, *2*(2), 121-136.

Hazell, P., & Haggblade, S. (Eds.). (2010). Transforming the rural nonfarm economy: Opportunities and threats in the developing world. *The Johns Hopkins University Press.*

Joshi, S., Singh, R. K., & Sharma, M. (2020). Sustainable agri-food supply chain practices: Few empirical evidences from a developing economy. *Global Business Review*, 0972150920907014.

Kayumova, M. (2017). The role of ICT regulations in Agribusiness and rural development.

Kishore, N. K. (2012). Agriculture credit in India: An integrated rural credit approach. *International Journal of Scientific and Research Publications*, *2*(3), 1-5.

Lutz, E., & Herrera, J. (2018). *Agribusiness for Africa's prosperity*. The World Bank. Retrieved from https://openknowledge.worldbank.org/bitstream/handle/10986/29944/Agribusiness0for0Africa0s0prosperity.pdf

Prabakaran, V., & Muthukumaran, C. K. (2020). The Rural Development Planning in Indian Economy-A New Century. *International Journal of Management (IJM)*, *11*(12), 2856-2863.

Rao, D. K. (2014). Impact of Food Security on Agri-Business in India. Retrieved from https://www.nssa.us/journals/pdf/NSSA_Proceedings_2014_Las_Vegas_1.pdf#page=202

Rao, N. H. (2007). A framework for implementing information and communication technologies in agricultural development in India. *Technological Forecasting and Social Change*, *74*(4), 491-518.

Selvaraj, M., & Ibrahim, M. S. (2012). Indian agricultural marketing-A review. *Asian Journal of Agriculture and Rural Development*, *2*(1), 69-75.

Sharma, S., & Tyagi, A. K. (2016). Integrating Corporate Social Responsibility Initiatives of Agro-Based Industrieswith Sustainable Rural Development in India". *International Journal in Management & Social Science*, *4*(6), 781-791.

Srivastava, P., & Agrawal, B. (2017). Innovative strategies for the development of rural India through village cottage and agribusiness enterprises. *Global Business and Economics Review, 19*(3), 276-298.

Swinnen, J. F., & Maertens, M. (Eds.). (2019). The Handbook of Agricultural and Rural Development in Emerging Economies. *Edward Elgar Publishing.*

Williams, S., & Karen, R. (2019). *Agribusiness and the small-scale farmer: A dynamic partnership for development.* Routledge.

World Bank. (2012). Gender Equality and Development. *World Development Report 2012*: The World Bank. Retrieved from https://openknowledge.worldbank.org/bitstream/handle/10986/4391/9780821394527.pdf

15

Innovative and Entrepreneurship in Agribusiness

Th Onchoila Maring[1], Pankaj Kumar[2], Raplang Lapasam[3], Maibam Suraj Singh[4]

[1]Ph.D. Scholar, Department of Agricultural Extension, School of Social Sciences, College of Post Graduate Studies in Agricultural Sciences, CAU (I), Meghalaya.
[2]Ph.D. Scholar, Department of Agricultural Extension, School of Social Sciences, College of Post Graduate Studies in Agricultural Sciences, CAU (I), Meghalaya
[3]Ph.D. Scholar, Department of Agricultural Economics, School of Social Sciences, College of Post Graduate Studies in Agricultural Sciences, CAU (I), Meghalaya
[4]M.Sc. Scholar, Department of Agricultural Extension, School of Social Sciences, College of Post Graduate Studies in Agricultural Sciences, CAU (I), Meghalaya

ABSTRACT

Innovation and entrepreneurship's combination has propelled transformational change in a variety of industries. Innovative entrepreneurship is the process of discovering opportunities and taking advantage of them by coming up with novel, disruptive ideas. It entails creating brand-new goods, services, and business strategies to meet the changing demands of the market. Individuals and organizations can promote economic growth, generate job opportunities, and tackle societal issues by fusing innovation and entrepreneurship. It also necessitates getting beyond obstacles including limited resource access, unpredictable markets, and regulatory restrictions. Collaboration, flexibility, and a continuous search for new opportunities are essential components of successful innovative entrepreneurship.

Keywords: Innovation, innovative, entrepreneurship, market, opportunities

Introduction

For nations aiming for sustained growth and development in a knowledge-based economy, fostering an entrepreneurial and innovative culture has become crucial. With its plentiful people resources, vibrant startup ecosystem, and quick technology breakthroughs, India has become a key player in the innovation sector. The idea of innovation has expanded over the past three decades to include not just the generation of new ideas but also acting as a catalyst for societal progression, technological development, and business success. Today, innovation provides a comprehensive response to a variety of problems. In the 20th and 21st centuries, unemployment has become a pervasive socioeconomic problem, refocusing attention on the crucial role that small firms play in job generation. The last ten years have seen a tremendous shift in the political economy of the world, which has heightened governments' interest in startups and small businesses. Rarely has there been greater consensus among nations on the need to develop or developed nations; everyone agrees that the private sector expansion is the key strategy for raising income. Nowadays, people look up to entrepreneurs as role models, and entrepreneurship education—the science of founding and operating a business—is one of the industries with the quickest growth rates.

Definition and Concept of Innovation and Innovative

Innovation is the process of coming up with and putting into use new concepts, items, processes, or strategies that lead to noteworthy developments, changes, or improvements. It entails coming up with original and innovative solutions to issues, grabbing chances, and offering something improved or new that provides value or addresses issues already present.

OECD (1981) defines innovation as all those scientific, technical, commercial and financial steps essential for the successful development and marketing of new or improved manufactured products.

Whereas Godin (2008) defines innovation in different concepts which are as follows

» Innovation as something new (some actual thing, such a product, service, or piece of software

» Innovation as the act of doing, or the means of making, something new;

» Innovation can be defined as a process of change, a circumstance (environment)

for creating something new, an idea (concept), human capabilities for doing something new, and more.

» Innovation as a prerequisite (environment) for attempting anything novel;

» Innovation as idea (concept) of something new;

» Innovation as the launch of a new product.

When innovations consistently improve the consumer experience that the business /enterprise provide to its market, they are considered innovative. When a corporation consistently improves the consumer experience it provides to its market. The company becomes more distinctive from its rivals as a result.

Indicators of innovative Innovations

Innovation can be called as innovative if they have these following attributes

1. Disruptive

Products, methods, and services are considered innovative when they disrupt the established order in their particular industries. For these innovations to differentiate themselves from other businesses in the same sector, they must be compelling and offer distinctive value and cost differentiation.

2. Must add value to buyers

This innovation must be valued enough for buyers to buy them or beneficial enough to improve people's lives. Innovative innovation is focused on the client and have the best cost and timing. When an innovation improves the lives of the company's clients as well as itself, it is considered innovative. It transforms information and concepts into advantages that can ease human life.

3. Introduces changes

Innovative ideas can change current systems and processes for growth in addition to being completely new and different.

Definition and Concept of Entrepreneur and Entrepreneurship

The word entrepreneur is derived from a French word "entreprendre" which means "to Undertake" Peter F. Drucker defines an entrepreneur as someone who actively seeks out change, adapts to it, and seizes the opportunity it presents. Entrepreneurs use innovation as a specific technique to take advantage of change as a chance to launch a new product or service.

Tamanna (2015) defined an entrepreneur as someone who exercises initiative by organizing a venture to take benefit of an opportunity and, as the decision maker, decides what, how, and how much of a good or service will be produced and entrepreneur is usually a sole proprietor, a partner, or the one who owns the majority of shares in an incorporated venture.

Cole (1959) defined entrepreneurship as the purposeful activity of an individual, or a group of associated individuals which is undertaken to initiate, maintain or aggrandize profit by production or distribution of economic goods and services.

According to Khanka (1998) an individual who has an urge to do (or) create something new, organise production, undertake risk and handle the economic uncertainty involved in running enterprise is called entrepreneur. The set of such attributes the entrepreneur possess is called entrepreneurship.

Samwel (2003) viewed entrepreneurship as a function which seeks investment and production process by raising capital, arranging labour and raw materials, finding site, introducing new techniques and commodities and discovering new sources for the enterprises.

Therefore an entrepreneur is someone who is able to identify unmet needs, have the abilities to see and evaluate business opportunities, bring on new ideas or innovations and take risks where as Entrepreneurship has been conceptualized as a process or function in some literatures and as an act or activity in some other. Nevertheless, it is associated with an entrepreneur and what he does in the pursuit of his business. It is all about setting up an enterprise taking on financial risks in the hope of profit.

Competencies required for successful entrepreneurship and entrepreneurial traits

Three major area of competencies are classified for successful entrepreneurship

1. A body of Knowledge

2. A set of skills

3. A cluster of traits

Knowledge

Knowledge refers to the understanding, information, abilities, and expertise obtained via education, experience, or study. It can also be defined as a set or body of information that may be saved and retrieved at a later time Knowledge in the business world can take the shape of awareness of or familiarity with things like business opportunities, the market, the customers, technical matters etc.

Skill

Skill can be defined as the capacity to put information into practise, which can be learned or improved through practice. E.g. swimming, driving etc. Some of the skills required in context of business are engineering skills, marketing skill, computing skill, financial management skill, planning skill, leadership skill etc.

Traits

Traits have been described as the combination of unique traits or features that make an individual unique. Some of the entrepreneurial traits are hark working, self-confident, profit oriented, goal oriented, persistent, copes with failure, risk takers etc.

Knowledge, Skill and traits are equally important and interrelated as because even if someone starts out business with just knowledge and abilities, it's unlikely that they will last very long. Likewise a person with only knowledge and traits might not be off much help without technical skill. In a same manner a potential entrepreneurs with skills and entrepreneurial traits but lacking in knowledge could lead to failure of his/her firm, as we know that information is really essential for any kind of business to succeed and flourish.

Therefore three areas of these competencies are required for successful entrepreneurship.

Importants of Innovative and Entrepreneurship

Innovation and entrepreneurship are key drivers of economic growth, technological progress, and societal transformation. In this rapidly evolving and interconnected world, the ability to innovate and seize entrepreneurial opportunities is crucial for individuals, businesses, and economies.

1. Economic Growth and Job Creation: Innovation and entrepreneurship are fundamental to fostering economic growth and generating employment opportunities. Research has shown that innovative businesses tend to be more productive, competitive, and resilient, contributing significantly to GDP growth. According to a study by the Global Entrepreneurship Monitor, countries with higher rates of entrepreneurial activity experience higher economic growth. For instance, countries like the United States and Singapore have thrived due to their emphasis on innovation and entrepreneurship, leading to increased job creation and improved living standards.

2. Technological Advancement and Industry Disruption: Innovation and entrepreneurship drive technological advancements and disrupt industries, leading to transformative changes. Startups and entrepreneurs often introduce disruptive technologies, challenging established market players and traditional business models. Examples such as Uber, Airbnb, and Netflix demonstrate how innovative business models utilizing technology have revolutionized the transportation, accommodation, and entertainment sectors. These advancements not only improve efficiency and convenience but also create entirely new industries, generating wealth and spurring economic growth.

3. Societal Development and Solving Global Challenges: Innovation and entrepreneurship are essential for addressing global challenges and fostering societal development. Social entrepreneurs, for instance, apply innovative approaches to tackle pressing issues like poverty, healthcare access, and environmental sustainability. Organizations such as BRAC, Grameen Bank, and Tesla exemplify how entrepreneurial initiatives have led to transformative solutions and positive social impact. By driving innovation in sectors like healthcare, renewable energy, and education, entrepreneurship plays a critical role in advancing societal well-being.

4. Cultivating a Culture of Innovation and Risk-Taking: Innovation and entrepreneurship cultivate a culture that values creativity, risk-taking, and continuous improvement. Organizations that foster an innovative mindset and encourage entrepreneurial thinking are more likely to adapt to change, seize opportunities, and stay competitive. By promoting a culture of experimentation, learning, and embracing failure as a stepping stone to success, innovation and entrepreneurship empower individuals and organizations to drive meaningful change and achieve long-term growth.

5. Collaboration and Knowledge Exchange: Innovation and entrepreneurship thrive in environments that facilitate collaboration and knowledge exchange. Partnerships between academia, industry, and government entities can foster a vibrant ecosystem where research findings can be effectively translated into practical applications. Open innovation models, such as collaborative research projects and startup incubators, encourage the sharing of ideas, resources, and expertise, enabling the development of innovative solutions with a wider societal impact.

Drivers of Innovation and Entrepreneurship

Technology advances

Technology is not a necessary component of innovation, technology frequently aids in the process of bringing about change. According to Tidd, Bessant, and Pavitt (2005), Joseph Schumpeter, an economist from the eighteenth century, suggested that businesspeople utilise technological progress to their advantage, giving them a temporary financial advantage (monopoly profits).

According to Schumpeter, this step change or discontinuous innovation entails creative destruction. Doing things differently can be viewed as step change innovation. The invention will typically come after a stretch of market stability and cause significant disruption. Technical success, customer acceptance, and competitive response are likely to be subject to far higher volatility and unpredictability. However, other entrepreneurs will imitate this and, as a result, other innovations (incremental innovations) will emerge, resulting in many new ideas that chip away at the 'monopoly profits' until equilibrium is reached and the process starts all over again.

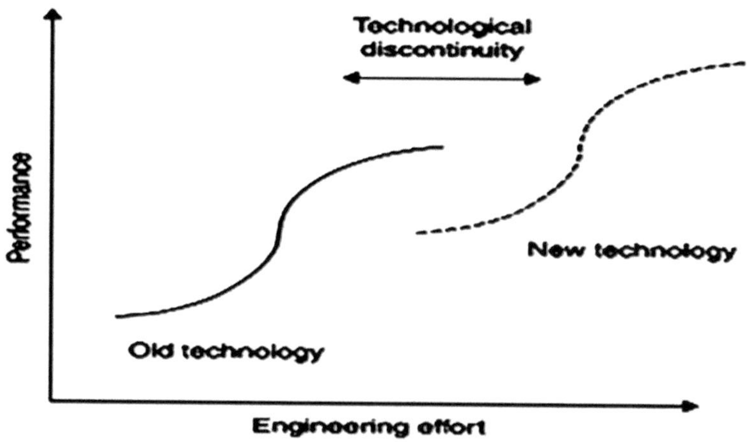

Fig.1: R&D effort and discontinuity

a. Technological discontinuities

Foster (1986) illustrated the 'S' curve of the technology life cycle, as shown in Figure 1. This graph shows how much money is spent on product development and customer-perceived performance enhancement. Investment in product, service and process development results in a stream of small performance improvements that add customer value. However, at the top of the 'S' curve, when the mature phase of the life cycle has been reached, even for quite large research and development investment made there is little further improvement in performance. For example, analogue TVs reached this point and even substantial additional investment could not achieve further small improvements in customer satisfaction. The price of digital TVs has decreased the price of older products, which has quickly decreased demand. This pattern follows a similar breakpoint and progression from black and white to colour televisions. You might make assumptions about what the next breakpoint might be in terms of efforts and resources used; for instance, it might involve completely new levels of interactivity or 3D imagery.

b. Technology life cycle

Technology evolution drives change within an industry sector but also provides the means for entrepreneurs to respond to change and build competitive advantage, as shown in Figure 2. Entrepreneurs are able to adapt and apply 'pure' science and technological inventions to create new commercial solutions in the form of new products, services and processes. Entrepreneurial businesses will gain a new source of competitive advantage if they successfully adopt a new technology and find a useful application, such as by developing a new product, service, or method of distribution. New standards are typically set for the industry sector as a result, and competitors must then meet these standards in order to continue to compete in the market. Then, every company in the industry adopts the new technology. In order to once again outperform the competition, the innovator firms must find a fresh innovation or technological advancement. This industry's cycle of technological advancement generates innovative new ideas that are not limited to the goods and services provided by the "for profits" sector.

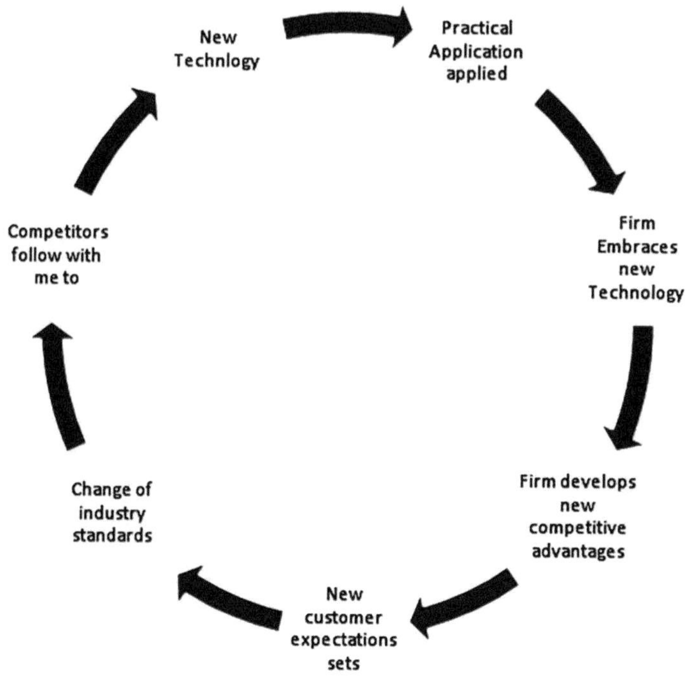

Fig.2: Technology and competitive advantages

Changing consumer preferences

Consumer preferences are expectations, likes, dislikes, motivations and inclinations that drive a customer to purchase specific products.A family might pick a restaurant because they enjoy a particular cuisine it offers. A customer buys something from a clothing store because of the available colour options or the discounted price tags. A news junkie might prefer The New York Times online over the printed version. Therefore, you can tailor your brand and products to be more appealing by using knowledge of your target consumer's preferences. You can gather customer preferences and respond to them with the aid of Possible NOW's technology and services. Business or entrepreneur's executives regularly conduct consumer research to spot shifts in customer expectations. You can create marketing campaigns and products that appeal to your company's target market and make your brand stand out from the competition by keeping up with shifting tastes and preferences.

a) Understand the changing preferences of customers

1. Segment your customers

Most businesses struggle to adapt to changing consumer preferences and neglect to use technology's advantages to better serve their clients. For instance, setting up an online store where customers from a distance can place orders and setting up doorstep delivery. Due to the influence of technology, consumer preferences are changing, and this includes the ways in which people shop and make payments. Knowing these variations will make sure that the company's customers are content and stick around for a long time.

2. Address customer complaints and issues

Making the time to hear and respond to customers' concerns and complaints is another surefire way to comprehend their shifting preferences. Bill Gates once said, "Your most dissatisfied customers are your greatest source of learning." All businesses would become much better at satisfying their customers if they realized this. When customers complain, it shows that they are attempting to offer feedback. In essence, they are giving the business a chance to improve. Companies that reject or ignore this "feedback" are essentially shutting the door to new business.

3. Conduct Research

There are many competitors in the market today. Each one is more assertive and capable than the one before it. Companies that are unable to adapt to the shifting consumer preferences will undoubtedly go out of business. Ask customers frequently if they are satisfied with the products and services the business offers. It enables a business to strengthen and improve what it already does well while addressing any weaknesses that client may have.

4. Companies Create Proper Marketing Strategies

The choice of messages and distribution channels is crucial because it will affect whether customers actually receive the messages. If so, whether or not those messages apply to them. Customers must be able to relate to the messages. They ought to be interesting and captivating. Customers are a wealth of information and perceptions. Businesses must make use of their knowledge to adapt their brands and products to customers' shifting tastes and preferences.

5. Companies Keep Their Existing Customers Happy

They anticipate that the business will consistently live up to and surpass their expectations. Understanding customers helps a business comprehend why a customer might leave or stop purchasing from them. The truth is that different customers define happiness and satisfaction differently. Customer satisfaction is therefore a complicated topic that businesses struggle to define and quantify.

Policy support

The majority of academics, decision-makers, journalists, and the general public assume that well-crafted government policies can encourage innovation in business practices and entrepreneurship. The overall effects of such policies, the precise mechanisms by which they function, and the boundary circumstances in which they apply, however, remain unclear. These important queries are frequently avoided by advocates, but they all demand thorough examination. For instance, are state-funded incubators and accelerators, public research parks, and other direct incentives like grants and cash subsidies more effective than indirect incentives like infrastructure support or technical training (Dimos & Pugh 2016; Lerner 2009). Is it preferable to concentrate on legal and regulatory policies instead, such as quick patenting, flexible labour laws, and fair business registration and licensing requirements.

The Indian government has implemented a number of programmes and policies to encourage an innovative and entrepreneurial culture throughout the nation. The Indian government has recently developed a wide range of new initiatives and opportunities in a number of industries to foster innovation.

In order to promote economic growth, job creation, and technological advancement, the government of India has made policy support for innovative entrepreneurship a top priority. To foster an environment that is favourable for startups and encourage innovation, the Indian government has put in place a number of initiatives and policies. Here are a few crucial policy areas:

1. **Startup India Initiative:** The Startup India Initiative, which was established in 2016, aims to support and promote startups in India. The registration process is made simpler, tax incentives are provided, and a Startup India Hub is established to give access to networking, mentoring, and funding opportunities. These are just a few of the advantages it offers.

2. **Tax Incentives:** To promote investment and innovation, the government provides tax incentives to startups. In their first ten years of business, startups

with annual sales of up to INR 25 crore (roughly $3.5 million) are eligible for a three-year tax holiday. The Angel Tax Exemption's implementation has also lessened the tax burden on startups.

3. **Fund of Funds**: A Fund of Funds for Startups (FFS) has been established by the government with a corpus of INR 10,000 crore (roughly $1.4 billion). FFS makes investments in Alternative Investment Funds (AIFs) that are registered with SEBI and then fund startups. By improving startups' access to capital, this initiative hopes to help them expand and scale.

4. **Protection of Intellectual Property Rights (IPR)**: The Indian government has taken measures to bolster the country's intellectual property ecosystem. To support and safeguard innovations, it has set up specialised IP facilitation centres and sped up the patent examination procedure. For the purpose of facilitating startups' intellectual property protection, programmes like the Scheme for Facilitating Startups Intellectual Property Protection (SIPP) offer financial assistance.

5. **Incubation and Innovation Centres**: To assist startups, the government has established incubation centres and innovation hubs across the nation. These facilities offer these facilities offer access to funding, mentorship, networking opportunities, and infrastructure. Atal Innovation Centres, Technology Business Incubators, and Atal Incubation Centres are a few examples.

6. **Business Ease**: The government has made it a priority to make doing business in India easier. Reforms have been made to streamline company registration procedures, lower compliance standards, and speed up clearance times. Different administrative processes for startups have been streamlined by the introduction of online portals and digital platforms.

7. **Skill Development and Entrepreneurship Training**: To foster a skilled workforce and encourage an entrepreneurial culture, the government has launched skill development programmes and entrepreneurship training. Aspiring entrepreneurs can receive training, mentoring, and financial assistance from programmes like the National Entrepreneurship Awards, Atal Innovation Mission, and Startup India Learning Programme.

8. **Public Procurement:** To encourage the participation of startups in public procurement, the government has introduced policies. It has set aside a number of product and service categories for startup companies only. As a result, startups are encouraged to create novel solutions and secure

government contracts.

9. **International Collaborations**: To encourage innovation and technology transfer, the government supports international collaborations. Partnerships, investments, and knowledge exchange between Indian startups and international competitors are facilitated by programmes like the India-UK Tech Partnership and the India-Israel Innovation Bridge.

10. **Support for Research and Development (R&D)**: The government encourages startups to invest in R&D projects. To encourage R&D and innovation among startups, a variety of grants, funding options, and incubation support are available.

Barriers to Innovation and Entrepreneurship

There are several barriers that can hinder innovation and entrepreneurship. These barriers can vary depending on the specific industry, country, and economic conditions, but here are some common ones:

1. **Financial barriers**: For many aspiring entrepreneurs, a major obstacle is a lack of capital and funding. Starting a new business requires a sizable investment, and getting financing can be difficult, particularly for those without a track record or collateral.

2. **Regulatory barriers**: Complex and onerous regulations can obstruct entrepreneurship and innovation. Startups may find it challenging to navigate the regulatory environment due to excessive bureaucracy, licensing requirements, and legal complexities, which can result in delays, higher costs, and legal risks.

3. **Intellectual Property Protection**: Insufficient or ineffective IP protection systems can stifle innovation. Entrepreneurs may be reluctant to commit time and money to creating new goods or services if they cannot adequately protect their inventions, ideas, or creative works.

4. **Market Barriers**: For entrepreneurs, breaking into a crowded or fiercely competitive market can be intimidating. It can be difficult for startups to gain market share and compete successfully because established companies with dominant market positions and economies of scale may have a significant advantage over new entrants.

5. **Limited Access to Networks**: For entrepreneurs, developing a strong network of advisors, investors, customers, and mentors is essential. However, a lack of connections and restricted access to networks can make it difficult for them to find assistance, forge alliances, and gain access to important resources.

6. **Risk aversion and Fear of Failure**: One of the main obstacles to entrepreneurship is the fear of failing. Starting a new business entails inherent risks, and people may be hesitant to take the plunge out of fear of suffering a financial setback or social rejection. Risk aversion can stifle creative thinking and restrict entrepreneurship.

7. **Cultural and social hindrances**: Social attitudes and cultural norms can affect how entrepreneurship is perceived. There might be a preference for stable employment or a stigma attached to failing businesses in some cultures. These elements may deter people from pursuing business opportunities.

8. **Lack of access** to programmes for entrepreneurship education and support can be a deterrent for would-be business owners. People may find it difficult to deal with the difficulties of starting and running a business without the necessary information, abilities, and guidance.

Case Studies of Successful Agricultural Innovations and Entrepreneurship

Vertical Farming

Despite the efforts of Green revolution, the growing population could not meet their needs of food production. So with this objective in mind, eminent researchers, scientist, agriculturist etc. started to look for various alternatives which could fulfill the requirements of human population. The term "Vertical Farming" coined by Gilbert Ellis Bailey came into existence in the year 1915. The modern concept of Vertical farming started with Professor Dickson Despommierin in the year 1999. Initially, Vertical farming means growing food in urban areas to meet the urban food needs. Now, Vertical farming means farming which involves the cultivation of plants in stacked layers. There are three systems of vertical farming namely- 1.Hydroponics (growing of plants in water), 2. Aeroponics (growing of plants in air using mist or nutrient solutions) and 3. Aquaponics (combination of hydroponics and aquaculture). This type of farming is gaining momentum in the 21st century as it could produce food in a sustainable way in an efficient manner (Sonawane, 2018) . There is lesser incidence of pests and diseases with maximum returns per unit area, no herbicides and pesticides requirement, less water requirement, more greeneries availability to the consumers with reduced supply chain and free from destruction by natural calamities. Although the initial cost of investment is high as it involves controlled environment like light, temperature, energy, air conditioning, ventilation system etc., this farming method is used widely in many countries owing to its maximum benefits. A number of companies has come up with this idea of farming and numerous entrepreneurs came into existence which are very successful in this field. USA is the country with highest number of vertical farms) while Japan being the highest in Asia (Vertical farming planet, 2023).

Some of the World's leading companies are as follows:

1. AgriCool Company in France came into existence in the year 2015 in France with the objective of producing food with no pesticides and preservatives. It produces fruit like strawberry, leafy vegetables like parsley, basil earning a revenue of US dollar 103 Million.

   Website: www.agricool.co
   Ownership status: Privately held
   Office: 139 Rue Rateau
   93120 La Courneuve
   France
   CEO: Guillaume Fourdinier

2. Sky Greens, the most famous vertical farming company in Singapore produces around 800 kg of vegetables daily. It is the world's first low carbon, hydraulic driven vertical farm (Sky Greens, 2023) established by Mr. Jack Ng with the objective of providing better agricultural solutions with lesser impact on natural resources through invention and innovation thereby providing food security to the people.

   Website: www.skygreens.com

For its great contribution, Sky Greens was awarded with Singapore Sustainability Awards 2014 by Singapore Business Federation, INDEX Award 2015 which is the world's biggest design award in the category of Work.

3. Bowery farm in US has the largest farming operations in Maryland, New Jersey and Pennsylvania cultivating pesticides free fruits and vegetable like lettuce, kale and basil earning a revenue of US dollar 83.7 million.

4. Spread Company in Japan established in 2006 is one of the Asia's leading vertical

farming company producing sweet strawberries and crispy lettuce with a daily production volume of 19 tonnes supplying about 5000 stores.

Precision Agriculture

Precision agriculture or site specific crop management is an integrated approach where an appropriate amount of inputs like fertilizers, water, light, temperature etc. are applied with respect to the requirements of the specific crop. It considers temporal and spatial variability for a sustainable crop production (ISPAG, 2021). Precision farming involves the integration of technology in farming practices thereby reduces the labour cost, saves time and timely action can be taken up for the crop to produce better yield. It involves using of GIS, GPS, Remote sensing, AI, Data collection and analytics, yield monitors etc. (AGRIVI,2023).

Some of the successful stories of farmers using Precision Agriculture are as follows:

1. Sultan Ameerden belonging to Velvarkottai Village, Dindigul district, Tamil Nadu is a Sugarcane farmer and he took up NADP assisted Precision Farming with the help of Agriculture Department staff. On 5th Dec, 2008, he planted CO-86032 variety and planted 30,000 two budded sets with a spacing of two and half feet row wise. He installed drip irrigation system with the help of the NADP scheme. For fertigation purpose, he used 55 Kg of NPK 19:19:19, MOP 37.5 Kg, Potassium Nitrate 65 Kg, S Potash 25 Kg. He could harvest 126.75 tonnes that fetched him Rs. 1, 90,125 gaining a profit of Rs. 1, 36,625 which was more than the profit he got in conventional method .He was happy with the experience he got through precision farming obtaining higher income and profit.

2. R. Velappan belonging to Semmanahalli Village, Dharmapuri district, Tamil Nadu installed drip irrigation in his 2 ha farm with Rs. 2,50,000 (75% subsidy), biofertilizer and vermin composite to use as manure. He planted papaya and guava producing an annual sales of Rs. 7,00,000 to Rs. 10,00,000. From his 2 acres land. He received training from KVK and TNAU about precision farming. He was satisfied that less usage of fertilizers reduces cost of cultivation thereby increasing yield with higher income.

Policy Recommendations for Promoting Innovation and Entrepreneurship In Agriculture

The government has placed entrepreneurship development in agriculture among its top priorities. This is a result of the fact that the majority of agricultural value chains have one or more problems that obstruct productive farming and the welfare of farmers. People are starting to recognise the importance of the agriculture sector for the growth of the country, which has led to a recent surge in new business ventures. It is supported by numerous organisations as part of their social responsibility. The Department of Science and Technology of the Indian government has started the process of establishing incubators inside of academic institutions.

Government of India for promoting innovation and entrepreneurship in agriculture implementing various programme and scheme which are as follows: -

1. A scheme for promoting innovation, rural industries and entrepreneurship (ASPIRE)

ASPIRE scheme is implemented by ministry of micro; small and medium enterprises (MOMSME) aim to support the incubation centres that accelerate entrepreneurship and promote startup in rural and agro- based industry.

The objectives of the scheme are to:

» To shape new jobs and alleviate unemployment,

» Promote the entrepreneurship culture in India,

» To support grassroots economic development at district level,

» To facilitate innovative businesses solutions for unmet social needs, and

» Promote innovation to further strengthen the competitiveness of the MSME sector.

2. Building awareness on intellectual property rights (IPR) for micro, small and medium enterprises

Building Intellectual Property Rights (IPR) Awareness for Micro, Small and Medium-sized Enterprises (MSMEs) is a programme that aims to i) improve the competitiveness of the MSME sector and ii) make the process of bringing new technologies to market easier.

» Objective of the Building awareness on intellectual property rights (IPR) for micro, small and medium enterprises

» The protection of concepts, technological advancements, and knowledge-driven business strategies created by MSMEs.

» The provision of suitable facilities and support for the protection and commercialization of intellectual property (IP) for the benefit of the MSME sector; and

» Aid MSMEs in making appropriate use of IPR instruments for technological advancement, market and business promotion, and improvement of competitiveness

3. Udyam Sakhi

An online platform called Udyam Sakhi, which is geared towards women entrepreneurs, gathers pertinent programmes on a single platform and provides access to entities that can help entrepreneurs as well as information on policies and programmes, financing options, and business ideas. Its goal is to help women launch, grow, and expand their enterprises in order to promote female independence.

4. National initiative for developing and harnessing innovations seed support system (NIDHI-SSS)

NIDHI-SSS offers financial aid to start-ups with promising ideas, innovations, and technologies, enabling chosen incubates to meet their capital requirements either through angel or venture funding or from commercial banks or financial institutions.

This scheme is implemented by Ministry of Science and Technology: Department of Science and Technology (DST).

Objective of this scheme

» Enable Science and Technology Entrepreneur Parks (STEP)/ Technology Business Incubators (TBIs) to expand their pipeline of start-ups and share their success, ensuring their long-term operational sustainability.

» Ensure timely availability of seed support for incubatees, enabling them to take their venture to the next level.

5. Technology Incubation and Development of Entrepreneurs (TIDE) scheme

Electronics and Information and Communications Technology (ICT) are the main topics of the "Technology Incubation and Development of Entrepreneurs (TIDE)" programme. In order to empower young entrepreneurs to launch technological start-ups for the commercial exploitation of innovations they develop, it aims to support higher education institutions in building their technological Incubation Centres (TICs).

The chosen companies are given working space by TIDE centres, which also offer nurturing support as needed. TIDE centres are typically established as societies, section 25 companies, or other types of legal entities by the Host Institute (HI).

The aim of the scheme is to:

» Set up or strengthen TBCs in higher educational institutions,

» Promote product-oriented R&D,

» Encourage and accelerate the development of indigenous products and packages, and

» Bridge the gap between R&D and commercialisation.

6. Atal Innovation Mission (AIM)

One of the flagship programmes of NITI Aayog is the Atal Innovation Mission (AIM). It is a key player in coordinating innovation policy among federal, state, and sectoral programmes that support innovation and entrepreneurship as an umbrella organisation and innovation centre. It works with MSME industries, corporations, NGOs, and higher education institutions as well as higher secondary schools, scientific, engineering, and academic institutions.

The objectives of the scheme are to:

» Serve as a platform for promotion of world-class innovation hubs, grand challenges, start-ups and other self-employment models, particularly in technology-driven areas, and

» Promote entrepreneurship through self-employment and talent utilisation.

Conclusion

Promoting innovation and entrepreneurship can have major economic and social benefits, and in order for firms to be competitive, they must use technology to their advantage and adapt to shifting consumer tastes because businesses that leverage technology, accept change, and adapt to shifting consumer tastes have an advantage in today's competitive market environment. India's suggested programme serves as an example of the country's dedication to encouraging and fostering creative enterprises and entrepreneurs in order to achieve these objectives.

References

Abhyankar, R. (2014). The government of India's role in promoting innovation through policy initiatives for entrepreneurship development. Technology Innovation Management Review, 4(8).

Accept mission. (2021). Innovation vs. Innovative vs. Creativity: The differences. https://www.acceptmission.com/blog/innovation-vs-innovative/#:~:text=Innovation %20refers%20to%20products%2C%20business,an%20approach%20than%20a%20result.%E2%80%99

AGRIVI. (2023). Precision farming: Everything you need to know. https://www.agrivi.com/blog/precision-farming/

Anon. (n.d). Introduction Entrepreneurship and Agribusiness Management. file:///C:/Users/USER/Desktop/Book%20chapter/Introduction/Enterpreneurship%20handout.pdf.

Anon. (n.d). The Nature and Importance of Innovation. file:///C:/Users/USER/Desktop/Book%20chapter/Introduction/nature%20and%20importance%20of%20innovation.pdf.

Bornstein, D., and Davis, S. (2010). Social entrepreneurship: What everyone needs to know®. Oxford University Press.

Bradley, S. W., Kim, P. H., Klein, P. G., McMullen, J. S., & Wennberg, K. (2021). Policy for innovative entrepreneurship: Institutions, interventions, and societal challenges. Strategic Entrepreneurship Journal, 15(2), 167-184.

Chesbrough, H. W. (2003). Open innovation: The new imperative for creating and profiting from technology.Harvard Business Press.

Christensen, C.M. (2013). The innovator's dilemma: when new technologies cause great firms to fail. Harvard Business Review Press.

Cole, A.H. (1959). The Tempo of Mercantile Life in Colonial America. *Business History Review*, *33*(3): 277-299.

Dimos, C., & Pugh, G. (2016). The effectiveness of R&D subsidies: A meta-regression analysis of the evaluation literature. Research Policy, 45(4), 797–815.

EduRev. (n.d).Policies of the Government for Entrepreneurs-Entrepreneurial

Sustainability. https://edurev.in/t/114191/Policies-of-the-Government-for-Entrepreneurs-Entre.

Emeritus, (2023). 5 Important Benefits of Innovation in Business, Blog. https://emeritus.org/in/learn/important-benefits-of-innovation-in-business/.

Godin, B. (2008). Innovation: The History of a Category. Sherbrooke Est Montréal, Québec Canada.

Hart, M., Bonner, K., Prashar, N., Ri, A., Levie, J., & Mwaura, S. (2020). Global Entrepreneurship Monitor:United Kingdom 2019 Monitoring Report.

ISPA. (2023). Precision Ag Definition. https://www.ispag.org/about/definition

Khanka, S.S. (1998). Making the entrepreneurial society. *Yojana*, 42(2): 9-11

Lerner, J. (2009). Boulevard of broken dreams: Why public efforts to boost entrepreneurship and venture capital have failed—And what to do about it. Princeton, NJ: Princeton University Press.

Lowe, R., & Marriott, S. (2006). Enterprise: entrepreneurship and innovation: concepts, contexts and commercialization. Routledge.

Maxim N. K., Alexander, S.A., & Meissner, D.(2013). Innovation Concepts and Typology – An Evolutionary Discussion. National Research University higher school of economics. file:///C:/Users/USER/Desktop/Book%20chapter/Introduction/innovation%20concept.pdf

MPRA. (2023). Practicing Precision Agriculture in Dharmapuri District of Tamil Nadu. https://mpra.ub.uni-muenchen.de/73233/1/MPRA_paper_73233.pdf.

PN. (n.d). Why Customer Preference Matters. https://resources.possiblenow.com/why-customer-preference-matters/

RKVY. (2023). Success is Sweeter than Sugarcane, through Precision Farming. https://rkvy.nic.in/static/download/RKVY_Sucess_Story/Gujarat/Sweeter_than_sugarcane,_through_precision_farming-TN.pdf.

Roberts, J.M., Bruce, T.J., Monaghan, J.M., Pope, T.W., Leather, S.R., & Beacham, A.M. (2020). Vertical farming systems bring new considerations for pest and disease management. Annals of Applied biology, 176(3), 226-232.

Samwel, K.L. (2003). Co-operative entrepreneurship – A perspective. *Kurukshetra,* pp. 20-23.

Shukla, U. (2023). Government Policies for Promotion of Entrepreneurship .https://www.tutorialspoint.com/government-policies-for-promotion-of-entrepreneurship#

Sonawane, M.S. (2018). Status of Vertical Farming in India. Int. Arch. App. Sci. Technol., 9 (4), 122-125.

Srinivas, R. Kumar, R., Gerard, M., & Srinivasa, C.R. (2019). Fostering Entrepreneurship in Agriculture through Incubation Centres in National Agricultural Research & Education System (NARES). Policy brief

Tamanna, R. (2015). Definitions of Entrepreneur.https://www.business management ideas.com /entrepreneur/definitions-ofentrepreneur-by-eminent-authors/19938.

Vertical farming planet (2023). Countries Using Vertical Farming – Vertical Farming Planet. https://verticalfarmingplanet.com.Accessed on 8th June, 2023.

16 Strategic Management: Formulation, Implementation and Control

Sachin Kumar and Arun Lather

[1]Department of Business management, College of Horticulture, Dr. YS Parmar University of Horticulture and Forestry, Nauni (Solan), Himachal Pradesh, India

Abstract

Any topic may be learned more effectively and is more fascinating if you have some understanding of how it has changed over time. This succinct part will cover the origins of strategic management and company policy before examining the discipline's historical progress and potential future developments. Finally, we'll look at some recent advancements in corporate strategy and policy in the context of India.

Keywords: Formulation, implementation Strategic control and Strategic management

Introduction

When Harvard Business School developed an integrated management course in 1911 with the goal of developing broad management competency, business policy had its beginnings. The case studies used in this course have been used at the institution since 1908 for educational reasons. Business policy is often regarded as a capstone, integrative course made available to students who have already taken a number of key functional area courses. The course's curriculum, teaching approach, etc. differ from institution to university. Although new titles like Strategic Management, Corporate Strategic and Policy, etc. are now widely used for the course, the word "business policy" has historically been utilised. The topic of the conversation up to this point was the academic standing of the business policy courses. However, in reality, progress has followed diverse trajectories.

According to Hofer and others, there have been four paradigm shifts in corporate policy development. These changes may be conveniently categorised as four overlapping periods in the growth of the topic of corporate policy. It's noteworthy to observe that the needs of real-lite business closely tracked the growth of business policy as a topic of study. The model of ad hoc policymaking, which Hofer claims dates back to the mid-1930s, was the foundation of the first phase. Due to the nature of American commercial enterprises at the time, policymaking became necessary. The businesses, which started off with a single product line serving a certain demographic of clients in a constrained geographic region, expanded in one or all of these three directions. With growth and the necessity to integrate the functional sectors, the informal control and coordination became only partially applicable. The formulation of policies to direct management activity led to this integration. Prior to subsequently taking on the job of senior management, former entrepreneurs were primarily responsible for formulating policy. Ad hoc policymaking was supplanted by systematic policy development in the United States throughout the 1930s and 1940s as a result of escalating environmental changes. Based on this second paradigm, the focus switched to the fusion of functional regions in a setting that was undergoing fast change. The planned policy paradigm was rendered obsolete by the environment's fast changes and rising complexity, which meant that policy-making and functional area integration were no longer sufficient to meet business demands. By the 1960s, there was a need for a critical examination of business's fundamental idea and its interaction with the environment. This condition was met by the notion of strategy, and the third phase, based on a paradigm for strategy, appeared in the early 1960s. The dynamic process of developing, implementing, evaluating, and controlling strategies to achieve the strategic objective of the organisation is referred to as strategic management. Strategic management is first and foremost a dynamic activity. This process is not static, one-off, or mechanical. Strategic management is a continuous, changing, iterative process because it is dynamic. As a result, strategic management cannot be a strict, stepwise collection of a select few actions put in a particular order. Instead, it is a dynamic mosaic of pertinent activity. Depending on the position they are in at the moment, managers may complete these tasks in any sequence. Iterative processes allow an action to be repeated as needed when the circumstances change rather than being limited to a single performance.

The formulation, execution, assessment, and control phases of the strategic management process are listed in the next section of the definition. There are four stages.

Four Phases in Strategic Management Process

- » Establishment of strategic intent
- » Implementation of strategies
- » Strategic evaluation
- » Formulation of strategies

Elements in Strategic Management Process

Each phase of the strategic management process consists of a number of elements, which are discrete and identifiable activities performed in logical and sequential steps. As many as twenty different elements could be identified in the models provided by various authors.

A. Establishing the hierarchy of strategic intent:
1. Creating and communicating a vision
2. Designing a mission statement
3. Defining the business
4. Adopting the business model
5. Setting objectives

B. Formulation of strategies:
6. Performing environmental appraisal
7. Doing organizational appraisal
8. Formulating corporate - level strategies
9. Formulating business - level strategies
10. Undertaking strategic analysis
11. Exercising strategic choice
12. Preparing strategic plan

C. Implementation of strategies:
13. Activating strategies
14. Designing the structure, systems and processes
15. Managing behavioral implementation
16. Managing functional implementation
17. Operationalizing strategies

D. Performing strategic evaluation and control:

18. Performing strategic evaluation
19. Exercising strategic control
20. Reformulating strategies

Model of Strategic Management Process

A model that comprises of several phases, each with a number of parts, is used to represent the strategic management process. The majority of writers concur that the strategic management process may be divided into four phases, each of which comprises roughly twenty aspects. The strategic management model that we use in this chapter is available.

Strategy Formulation

Environmental Appraisal

Literally, "environment" refers to the physical surrounds, outside factors, influences, or conditions under which something or someone lives. Any organization's environment may be defined as "the totality of all circumstances, occurrences, and influences that surround and have an impact on it." Understanding the environment is essential since it affects an organisation in so many different ways. By examining some of its properties, the notion of environment may be comprehended.

Characteristics of Environment

The business environment, or simply environment, demonstrates a variety of traits. Here, some of the crucial and evident traits are succinctly presented.

1. The environment is intricate. The environment is made up of a variety of elements, occurrences, circumstances, and effects that come from many sources. All of these interact with one another to produce completely different sets of impacts, rather than existing in isolation.

2. The environment is changing. In nature, the environment is ever-changing. The environment is dynamic and constantly changing in terms of shape and personality because of the numerous and diverse variables at play.

3. The environment is complex. Depending on how the observer sees things, an environment will take on a certain shape and personality.

4. The environment has a significant influence. Organisations are significantly

impacted by the environment. The environment in which an organisation operates has a significant impact on its growth and profitability. Any change in the environment has a variety of effects on the organisation.

External and Internal Environment

The variables outside the organisation that present possibilities or hazards to the organisation are all included in the external environment. The term "internal environment" refers to all internal elements that influence a company's strategic strengths or shortcomings. Therefore, in addition to the internal environment's strengths and weaknesses, the environment in which an organisation operates may be characterised in terms of the possibilities and risks present in the external environment. The following are possible descriptions of the four environmental influences:

1. An opportunity is a favourable circumstance in the surroundings of the organisation that allows it to solidify and improve its position. Growing demand for a business's goods or services is an example of an opportunity.

2. An unfavourable environment factor that puts the organisation at risk or harms it is referred to as a threat. A powerful new rival entering the market who is anticipated to provide fierce competition to the established businesses in an industry is an example of a threat.

3Strength is a natural ability that an organisation may employ to its advantage. Superior research and development abilities that may be applied to the development of new products to provide the firm a competitive edge are an example of a strength.

4. A weakness is a built-in restriction or limitation that results in tactical disadvantages. Overdependence on one product line is an illustration of a vulnerability that might be harmful for a business during times of crisis.

Organizational Appraisal

Different functional domains that are crucial for formulating strategy inside an organisation. For the purpose of identifying the firm's main strengths (and weaknesses) so that it can take advantage of opportunities and counter threats in the environment, strategists assess the firm's resources and capabilities in the key functional areas. Analysis of the following internal elements is the foundation for determining a firm's strengths and weaknesses:

Financial Capability

The availability, use, and administration of finances, as well as any other related issues, are all related to an organization's capacity or ability to carry out its strategies. The following methods can be used to determine the regions' strengths and weaknesses in accounting and finance.

1. **Financial resources and strength:** It aids in determining the source of funding and the availability of financial resources.
2. **Capital structure:** The capital structure of an organisation influences its capacity for flexibility in capital raising, financial leverage, and capital cost minimization.
3. **Financial planning:** Financial planning is the process of determining in advance the amount and kind of capital requirements. The organisation will be stronger if all of these things are planned out well in advance.
4. **Accounting system and audit procedure:** The effectiveness of accounting procedures and systems for costing, budgeting, profit planning, and auditing are determined. Determine if there has been no theft of funds and offer suggestions for further steps.
5. **Tax planning and Tax advantage:** Does the business benefit from tax benefits and use tax breaks obtained via tax planning.
6. **Relations with shareholders:** Do the company's policies on dividends and profit retention meet the expectations of shareholders? If this kind of connection is friendly, the business can continue to operate efficiently even in the face of difficulty.

Marketing Capability

The importance of marketing and distribution aspects may be attributed to the fact that successful company operations depend on them. The pricing, marketing, and distribution of goods and services, as well as any other related elements, are considered capability factors since they affect an organization's capacity and ability to carry out its strategies. The marketing function is regarded as an important sector because it not only affects the success or failure of corporate operations, but also because it serves as a crucial communication and interaction channel between the organisation and the outside world. The following components should be included in the study of marketing capabilities:

1. **Competitive position and market share:** Business organisations must function in a cutthroat environment. Therefore, it is important to understand whether the company has a substantial market share.

2. **Product line:** Are there many different styles and quality available in the product line?

3. **Product life cycle:** Knowing whether phase the primary products are in—introduction, growth, or declining phase—helps with product life cycle.

4. **Pricing:** To determine whether the price plan is successful, one must examine pricing policies, revisions, protection, etc.

5. **Market:** What kind of market is that? Do a select few consumers determine the sales?

6. **Marketing research:** Information from marketing research may be used to make decisions about pricing, packaging, advertising, and other aspects of marketing.

7. **Channel of distribution:** The term "channel of distribution" refers to the method by which products will be sold, including whether they will be sold through department stores, wholesalers, or other merchants. A strong distribution channel helps the company since it gets the goods to the right places when it's required, yet a poor channel can be one that's centralised..

8. **Packaging:** How well-performing and efficient are the packaging and related services.

9. **Marketing policy:** Is the policy in line with customer desires, technical advancements, and other environmental considerations such as competition?

Operations Capability

The production of goods and services, the use of material resources, and all other related issues that affect an organization's capacity and capacity to carry out its plans are referred to as operations capability factors.

1. **Factors related to the production system:** A few examples of these factors are capacity, location, layout, product or service design, work methods, level of automation, and vertical integration.

2. **Factors related to operation and control system:** Production planning in general,

material sourcing, inventory management, cost and quality control, maintenance systems and practises, etc.

3. **Factors related to R & D system:** Product development, patient rights, degree of technology employed, technical collaboration and assistance, etc. are only a few examples of the personnel facilities.

Personnel Capability

The availability and use of human resources and skills, as well as any other related elements, are all considered to be personnel capabilities factors that affect an organization's capacity and ability to carry out its strategies. The following are some significant variables that might affect an organization's human resource capacity:

1. **Factors related to the Personnel system:** System for managing human resources, including policies and standards, recruitment, training, remuneration, communication, and evaluation of the personnel department's role within the company.

2. **Organizational and employee's characteristics:** Corporate image, the calibre of management, staff, and employees, perceptions of the organization's reputation as an employer, the accessibility of possibilities for employee development, working conditions, etc.

3. **Industrial relations:** Employee satisfaction and morale, the union-management relationship, collective bargaining, safety, welfare, and security, etc. Is there a high level of unionisation among employees? Does the management keep cordial ties with the unions??

General Management Capability

General management capability refers to the coordination, integration, and leadership of the functional capabilities towards shared objectives and all other related factors that affect an organization's capacity and ability to carry out its strategy. Following are some significant aspects that affect an organization's overall management capacity:

1. **Management system:** Processes involved in strategic management systems include mission-purpose and objective setting, strategy assessment systems, management information systems, corporate planning systems, and systems enabling senior managers to be rewarded and incentivized, among others.

2. **External relationships:** Influence over and a good working relationship with the

government, regulatory bodies, and financial institutions; public relations; a feeling of social responsibility; a positive public perception of the company as a corporate citizen; etc.

Industry, Competitive and Internal Analysis

Industry

An "industry" is made up of a collection of businesses that "provide products or services that are near equivalents for one another. The goods meet the same fundamental requirements of customers. An industry is a collection of businesses whose goods have so many characteristics that they are in direct competition for the same customers. In contrast, an industry is a collection of businesses that provide identical goods or services to clients. For instance, "Kurkure," "Bingo," or "Lays Chips" are part of a "industry" because they meet Indian customers' demand for "mid-day hunger." Additionally, they are good alternatives to one another.

Purpose of Industry Analysis

The business operations of a corporation are greatly influenced by the industry environment. A corporation must thus make sure that its strategy matches the environment of its industry. If it becomes extremely challenging or impossible to do, the business must change the climate of the industry to its benefit by implementing the right strategy. Without a thorough examination of the industrial environment, no organisation can expect successful strategy-making. Because of this, it is commonly accepted that thorough industry and competition study should come before developing a strong strategy. Information on the circumstances facing the sector is provided through industry analysis. Understanding the industrial environment's function is a good place to start when analysing it. Understanding the competitive interactions between groupings of businesses that compete for a certain market is the goal of researching the industrial environment or analysing industry structure.

Generic strategies

The corporate level generic strategy have to do with choosing the industries that the organisation will operate in. They control the strategy that a business uses to achieve its goals. A company may be a simple, one-business operation or it may be a large, complicated, diversified operation with a number of enterprises. In all cases, the corporate strategy is closely tied to the overall direction of the business. For a small business, it may determine the course of action that would increase profits. The corporate strategy for a large company refers to managing the many companies to maximise their contribution to the achievement of overarching corporate goals. Alternative paths of action might be thought of in one of four general ways, according to Glueck: stability, expansion, retrenchment, or combination. On sometimes, the term "grand strategy" is used to refer to these broad tactics. Business enterprises research the general strategy possibilities while creating their corporate strategy since only via this process can they determine the precise path most advantageous for achieving the chosen growth aim.

Corporate strategy is anxious about two main issues:

(1) What business domains a company be engaged in so as to make its long-term profit as large as possible?

(2) What strategies should it apply to become involved in and withdraw from business domains?

Corporate strategies are essentially commitments to allocating resources to different businesses within a company, reallocating resources from one class of businesses to another, and managing a variety of investments in order to successfully achieve the broad corporate objectives. An organisation can carefully consider a variety of strategic choices thanks to an analysis based on business definition.

Stability strategy

Stability is the most advantageous technique to employ when a firm thinks that it must continue in its current business and is doing quite well in that business but lacks significant room for expansion. Jauch and Glueck defined a stability strategy as a strategy that a firm proceeds along when:

1. It persists to serve the customers in the existing product or service, market, and function areas as described in its business definition, or in identical sectors.

2. Its key strategic decisions concentrate on improvement that brings small change in functional execution.

The stability approach does not include taking no action. There are minimal improvements involved. Reinvestment, R&D, and innovation are also required for a long-term stability plan. The business definition, however, has not altered.

Expansion Strategy

When a company delivers public services in additional product or service sectors, or when it adds new markets or functions to its business definition, it is using an expansion strategy. It focuses its strategic efforts on accelerating activities within the parameters of its current business description. The business must be defined differently for this approach, either by broadening its range of activities or by significantly

Combination Strategy

A company is said to be pursuing the combination generic strategy when it combines stability, expansion, and retrenchment either concurrently or sequentially to improve its performance. Combination strategies are when the strategists purposefully use distinct general strategies to diverse areas of the company or to various future timeframes. A firm may pursue stability in certain companies while expanding in others, or stability in some businesses while cutting back on staff. These are the justifications for adopting a variety of strategies at the same time.

Retrenchment Strategies

When a company intends to drastically reduce the scope of its operations, the retrenchment generic strategy is the one that is chosen to be implemented. In order to achieve this goal, the issue regions and their root causes are recognised. After then, efforts are undertaken to find solutions to the issues that lead to various retrenchment techniques.

A retrenchment strategy can take any of the following forms:

1. Turnaround strategy
2. Divestment or divestiture strategy
3. Liquidation strategy

Turnaround Strategies

When a company is steadily losing money but yet valuable to save, turnaround tactics are helpful. Recognising the circumstances in which an otherwise successful company could falter is critical. Such recognition can offer the suitable fixes that can promote recuperation.

Divestment Strategies

Regular business divestitures—even of some profitable, strong ones—ensure that the remaining divisions reach their maximum potential and that the corporation as a whole becomes more powerful. Some business leaders understand the significance of a professionally planned divestment programme. There are a number of expenses associated with holding onto a firm too long.

» Costs to the corporation
» Costs to the unit
» Depressed exit price

Though these costs are concealed, they can be greater than the benefits of holding the business.

Costs to the corporation

On the one hand, well-established, prosperous companies may create money and offer steady profits. On the other side, they can cripple a business to launch brand-new, high-growth ventures. Businesses that are committed to development frequently need to feel as though they are in a crisis. A firm may receive the funds to thrive today from long-held, low-growth enterprises, but they may prevent it from setting itself up for a prosperous future.

Costs of the Unit

When a business unit is maintained for an excessively long time, both the unit and the organisation as a whole suffer.

Depressed exit price

A divestiture that takes place at a good moment can increase shareholder value, whereas one that happens at a bad time might decrease value. Many businesses don't unpack a unit until after a number of years of subpar performance. Some sectors are so turbulent that managers are unable to anticipate market ups and downs. In other situations, they could be able to spot the peaks but unable to locate a buyer willing to pay the going rate. A sale made sooner would have generated much greater profits for the vast majority of divestitures. Foster and Kaplan contend that a company's performance for shareholders degrades with time.

Liquidation Strategies

Liquidation strategy, which calls for closing down a business and selling its assets, is one of the most extreme and unwelcoming retrenchment tactics. Because it results in major consequences like job loss for workers and other employees, the end of chances where a company may pursue any more operations, and the stigma of failure, it is seen as the final resort. Medium- and large-sized businesses seldom liquidate in India, in contrast to the numerous small-scale units, proprietary businesses, and partnership ventures that do so often. The management of the firm, the government, banks and other financial institutions, trade unions, suppliers and creditors, as well as other authorities, are incredibly reluctant to decide or request liquidation. Due to the difficulty in finding purchasers, selling assets to implement a liquidation strategy may also be complicated.

Business Level Strategies

The process of business definition is visualized in terms of:

- Customer needs
- Customer groups
- Distinctive competencies

These choices are the foundation of business-level strategic choice since they determine a company's competitive edge over its rivals and how it will handle competition in a given market or sector. The business level tactics take into account the ways in which businesses might get a competitive edge.

CUSTOMER NEEDS

Customers' needs signify if a product's or service's qualities are up to par with their expectations. Basic requirements include the need for food, clothes, and housing as well as medical and higher order needs like the desire for security, knowledge, education, information, and social respect, among others. The satisfaction of these demands is the goal of business. Businesses work hard to understand client wants, satisfy them, and build products and services appropriately. In order to attract clients and satisfy a certain threshold of customer demands, all businesses must to some extent distinguish their products from one another. However, some businesses differentiate their goods far more than others, and this differentiation can provide them an edge over the competition. Some businesses choose to charge their clients a low price rather than differentiating their items.

Customer Groups

In order to gain a competitive edge, businesses identify client groups based on the key variations in their demands or preferences. This is market segmentation in action. Market segmentation enables businesses to provide a wide range of products for several market segments, better meeting the demands of their customers. As a result, the company's products are in more demand, generating more income than it would if it just sold one product to the market. Customers judge a product based on the price if the nature of the product or the industry does not afford a significant level of distinction. The firm with higher efficiency and the ability to offer a product at the lowest price has the competitive edge.

Distinctive Competencies

Identifying the particular skills pursued to satisfy client wants and groups is the key factor for the business level strategy. The manner in which a business tries to address the demands of different consumer groups in order to gain a competitive edge are known as distinctive capabilities. A company must decide how to organise and combine its unique skills in order to get a competitive edge when choosing its business strategy.

Strategic Analysis: Corporate, Business and Industry

Strategic Choice Process

The process of making a strategic decision essentially involves making a decision after formulating goals, coming up with alternatives, choosing one or more options that will help the business achieve its goals in the best possible way, and then putting the chosen option into practise. A decision-maker must establish an assessment criterion to rate each option's quality before choosing one among the many possibilities. The criteria for selection serve as a guide for decision-making and significantly simplify the selection process. The strategic choosing process consists of these four steps:

» Paying particular attention to alternatives

» The selection factors

» Assessment of strategic alternatives

» The strategic choice

Selection Factors

The process of narrowing down the strategic decision to a few workable alternatives to be further analysed is facilitated by the business description and a well executed gap analysis. In order to develop the test for the evaluation of strategic options, this study depends on a few selection parameters. The selection criteria are primarily divided into subjective and objective criteria. In order to facilitate a strategic decision, the objective elements are founded on analytical methods. The subjective elements are dependent on individual perception and judgement. Both sorts of selection variables are used to analyse the alternatives generated in the preceding stage.

Assessment of Strategic Alternatives

The final strategy choice is based on the selection criteria. Every individual strategic option has to be evaluated for its potential to help the organisation achieve its goals. Evaluation of strategic options necessitates thorough consideration of the findings of analyses based on both objective and subjective variables. Strategists are free to use any tactic that is appropriate for the situation.

Strategic Choice

An evaluation of strategic options leads to an evaluation of the best option. It is necessary to create a strategic plan outlining the strategies and the prerequisites for their effective implementation. In addition to the chosen techniques, certain backup plans would need to be created. The strategic decision-making process is carried out on two levels: the corporate level and the company level.

Strategic Choice at the Corporate Level

Strategic analysis is concerned with a corporate body made up of a portfolio of firms at the corporate level. The examination meticulously considers all of the many problems involving the various corporate portfolio companies. The general tactics of stability, expansion, contraction, and combination make up the available strategic possibilities. A firm with many businesses might benefit from a corporate level strategic study. company-level strategic analysis would be sufficient for lone company entities.

Selection Factors

Four managerial selection factors influence the strategic choice.

- » Perception of external dependence,
- » Attitudes toward risk,

» Awareness of past enterprise strategies

» Power relationships

Perception of external dependence

The other components on which commercial enterprises rely for their existence and prosperity are the owners, rivals, clients, authorities, and the community. A company's strategic choice is less adaptable the more dependent it is on these components. The variety of strategic options is so constrained. Interactions between the company and its surroundings lead to strategic decisions. Therefore, while different parties work to achieve their goals, strategic decisions are the outcomes that are discussed.

Managerial attitudes toward risk

The level of risk that the company, its investors, and management are willing to accept has an impact on the strategic option. The risk-taking attitude of managers can range from accepting risk to being strongly risk-averse. The risk-averse individuals will likely choose only defensive measures with extremely minimal risk since they believe the company to be very weak. Risk attitudes can fluctuate and vary by industry, which is prone to fast change, and by environmental unpredictability. Executives must be able to accept increased levels of risk in sectors that are subject to fast change; otherwise, it will be challenging for them to perform.

Managerial Awareness of Past Strategies

The origin of a strategic decision is a previous strategy, which may afterwards entirely eliminate some strategic alternatives. The starting point of the procedure is the firm's current status. At this stage, it is important to consider if sticking with the current course of action will result in the accomplishment of the intended goals. The old technique will be applied as long as the deficit is minimal. Additionally, managers will not consider other options if they are determined to sticking with the current course of action.

Managerial Power Relationships

People are aware that power dynamics play a major role in organisational life. In many businesses, the choice is made unanimously if the senior management supports it. The process of strategic decision-making occasionally involves personalities: The people the boss respects and likes have a significant impact on the decision-making process. Additionally, if "mistakes" are made, the balance of power may shift to lower-level leaders. Additionally, the CEO's influence matters. Personal objectives, aspirations, values, and motivation of the manager also have an impact on the strategy chosen. With the existence of a powerful CEO the organization's goals become interwoven with individual goals in the process of strategic chosen. When an organisation has a strong CEO, individual ambitions and organisational goals are intertwined during the strategic decision-making process. Politics and power undoubtedly also affect strategic choices. The relative merits of analytical, political, and intuitive methods to decision making depend on the significance of the decisions, the degree of time constraint and uncertainty, and the decision maker's personality. The compromise between the aims is also influenced by the political influences from outside. Politics is always at play, even to the point of influencing goals and how analytic methods are used and perceived. Politics appears to be the deciding element in the process of making strategic decisions around 30% of the time, according to (Mintzberg). As a result, it is also necessary to analyse the key managers' values and objectives. If the planned strategy is not carried out well, its chances of success may be low. It is also unlikely that a politically unpopular technique would be effectively eradicated.

Selection of Strategies

The decision on which strategies to choose for execution is not made by simply evaluating strategic alternatives. Johnson and Scholes provide three methods for choosing tactics.

Selection against objective

The organization's quantifiable objectives serve as the initial criterion for choosing future tactics. The decision-making process is consequently centred on evaluation methodologies. They should offer quantitative evaluations of the relative advantages of various options and suggest the best course of action. The objectives and the approach occasionally need to be adjusted. The strategist must determine if the strategic alternatives align with the organization's stated objectives.

Strategic Choice

The capacity to identify the most pressing challenges and to make wise decisions about how to deal with them is at the core of successful strategic planning. The decision to select the strategy that will best help the organisation achieve its goals from a variety of options is referred to as a strategic choice. An organisation is better prepared to make long-lasting strategic decisions with an efficient strategic choice process. When making a strategic decision, it is crucial to keep both the internal and external contexts in mind. Stability, growth, retrenchment, a combination, and any tactical variant of these general tactics are potential options.

Attributes of Strategic Choice

The following qualities define a solid option in the context of planning for an organization's long-term future performance:

Authenticity:

A strategic choice must be chosen from many, preferably at least two, possibilities that are realistic alternatives in order to be considered legitimate. The decision or choice must be made in such a way that it is plain both what the organisation will do if it chooses one of the possibilities and, more importantly, what it will not do. The entity must decide whoever it wants to use for its own good. It must decide which advantage it will offer. It must be obvious to a company whose clients it will serve and which markets it will not. Choosing the foundation on which a company will compete, or how a company will gain a competitive edge in the chosen client segments of the market, as well as what a company will not do to compete, are examples of authentic choices.

Achievability/Believability:

A believable decision is one that can be linked to some genuine foundation in terms of the proof and commitments of the team members in charge of the strategic planning. Choices that are believable are supported by reliable, representative, and pertinent evidence or are inferred from such facts by competent reasoning. The participants' intuition, emotional attachments, and personal ideas and values can nonetheless play a part in decisions, despite the emphasis on logic and evidence. However, it does recognise them and any potential impact they may have on the process of making strategic decisions.

Communicability/Understandability

However, if a strategic decision is not presented in a way that makes it possible for it to be successfully and efficiently conveyed, it is a waste of time. "Message received but not comprehended." This is not only a question of how it is said or written. It also depends on how tempting the option is. Any decision taken should be interesting to a range of audiences when conveyed through various planning processes as well as the intricate webs of formal and informal communication channels needed to get anything done in an organisation. These audiences may include those who were not directly involved in the formulation of the strategic decisions. However, it may be expected of these same individuals to carry out the related decisions or at the very least adapt to their effects.

The management team's dedication is put to the test in terms of the strategic decision's communicability. The manner in which the top managers convey the strategic option or choices will be used by members of the organisation who are not directly involved in the decision-making process but are nevertheless responsible for carrying out the strategic plan to assess the management team's enthusiasm. Naturally, they will compare the strategic decision's reasoning to their own experiences, and if they find it lacking, the strategic choice will not have been completely communicated since the recipients may reject it. "Message received, comprehended, and rejected."

Implementability

Exploring options related to resolving these challenges and exposing the strategic elephants or truly serious concerns is all very well. It's OK to reduce the options to a manageable number that satisfy the requirements A, B, and C. A strategic decision, though, isn't much value if it can't be put into practice. The decision is therefore not just Authentic, Believable, and Communicable. Additionally, it must be deliverable, doable, or...executable. The strategic options that have been shortlisted for inclusion in a company strategy must be able to be divided into a number of immediately actionable steps and then further divided into feasible medium- and long-term actions with clearly defined deliverables. These more significant corporate strategy decisions must also be able to be implemented in the actual world and be authentic, believable, and communicative. They need to be able to be converted into concise budgets. action plans or projects with well-defined accountabilities. They need to have at least sufficient resources. They must be designed with corresponding risk management procedures in place in order to cope with the inevitable roadblocks that may appear, which is equally crucial.

Strategy Implementation

The process of turning a strategy or plan into action is referred to as strategy implementation. It is the culmination of all the actions and choices necessary to properly carry out a strategic strategy. The process by which an organisation should create, use, and combine its organisational structure, control systems, and culture in order to implement strategies that provide it a competitive edge and improve performance is known as strategy implementation.

Role of Strategists in Strategy Implementation

The Board of Directors is only frequently involved when significant organisational changes are necessary, such as when a new CEO must be appointed, when significant structural changes must be made, or when significant resource allocation choices must be made in the event of mergers or capital investments. Top corporate level managers frequently participate in goal-setting and strategy creation, and they are crucial in fusing the corporate level strategy with the SBU level plans. With SBU managers, they bargain frequently over resource distribution. As a result, the distribution of resources is essentially the responsibility of top corporate managers and heads of SBUs. After fundamental choices are taken, a strategy is put into action by the organisational hierarchy. The senior executives of SBUs are informed of decisions/options taken at corporate headquarters. The SBU's senior managers then decide which precise techniques to employ in their organisations. The procedure is facilitated by the corporate planning team. For professional advice, consultants may also be commissioned.

Challenges in Strategy Implementation

The process of implementing a strategy is not as easy and straightforward as some decision makers may believe. The individuals in charge of carrying out plans may make decisions based on their perceptions of the situation, their prejudices, and their preferences. For instance, a new manager may apply a procedure that was successful in his previous organisation, or the CEO might introduce a brand-new implementation model that all employees are required to use. Consultants with their own tools and philosophies may step in to help. Complex theoretical models may excite certain managers, or someone may feel inspired to develop something entirely new. Most supervisors are unable to comprehend the procedure. Your strategy may be excellent, and your preparations for implementing it may be flawless, but if your management don't understand it, it is useless.

Ownership of the strategy implementation process is frequently unclear. There are numerous distinct supervisors handling the duties. In big organisations, the finance department, human resources, the strategy coordination team, internal consultants, various programme management offices, and last but not least, the managers themselves, may be given ownership and accountability for execution. Due to this dispersion, the global implementation process lacks ownership.

Evaluation of Strategy

Strategic evaluation is an assessment process that gives executives and managers information on how well projects, initiatives, and other activities are meeting their organization's goals and objectives. Of course, the decision-makers want to guarantee that the strategic option is carried out correctly. Due to its ability to shed light on the success and efficiency of comprehensive plans in reaching the desired outcomes, strategy evaluation is just as important as strategy design. The creation of new strategic planning, the requirement for feedback, evaluation, and reward, the development of the strategic management process, determining the viability of strategic decision, etc. are just a few of the reasons why strategic evaluation is important.

The procedure for evaluating a strategy entails:

1. **Establish Performance Targets, standards and tolerance limits-** Strategists must consider issues including what benchmarks to set, how to create them, and how to communicate them when establishing the benchmark. It is crucial to learn about the unique criteria for carrying out the primary work in order to establish the benchmark performance to be used. It is therefore possible to choose the performance indicator, which will be used for evaluation, that best identifies and expresses the unique needs. The organisation can evaluate performance using both quantitative and qualitative factors. Calculating net profit, ROI, earnings per share, production costs, staff turnover rates, and other quantitative metrics are among the quantitative criteria. Subjective assessments of variables like skills and abilities, risk-taking propensity, adaptability, etc. are included in the qualitative elements.

2. **Measurement of performance** - The benchmark against which the actual performance is to be compared is the standard performance. Performance management is made easier by the reporting and communication tools. Evaluation of a plan is made simpler if the relevant criteria are specified and the right tools are available for monitoring performance. However, other aspects, like the influence of managers, are challenging to quantify. Similar to individual performance, divisional performance can also be challenging to gauge. As a result, flexible goals must be developed in order to assess performance. The assessment will not serve its goal if the measurement is not performed at the appropriate time. Financial statements like the balance sheet and profit and loss account must be created annually in order to measure performance.

3. **Analyzing Deviations** - There may be variations that need to be analysed while measuring the real performance and comparing it to the standard performance.

The degree of tolerance limitations between which the difference between actual and expected performance may be acceptable must be mentioned by the strategists. Positive deviation denotes higher performance, although consistently surpassing the objective is highly exceptional. Because it shows a performance gap, the negative deviation is a cause for concern. Therefore, in this scenario, the strategists must identify the root causes of deviation and implement corrective measures to eliminate them.

4. **Execute modifications and refinements/Take Corrective Action** - It is critical to design a remedial action as soon as the performance discrepancy is discovered. If the performance repeatedly falls short of the intended performance, the strategists must do a thorough investigation of the contributing variables. The criteria must be decreased if the strategists find that organisational potential does not fulfil performance needs. Reformulating the strategy is another uncommon and harsh corrective measure that necessitates returning to the strategic management process, reshaping plans in accordance with new resource allocation trends, and as a result necessitates returning to the strategic management process' starting point.

Criteria for Evaluation

Both objective and subjective criteria may be used to establish evaluations. The evaluation's goal will determine the criteria to be employed. For the aim of evaluation, quantitative or qualitative methodologies may be utilised.

Quantitative Criteria: The firm's present performance is compared to its historical performance or that of its rivals in order to assess the success of strategy. This comparison includes metrics like earnings, dividends, stock prices, return on capital, market share, sales growth, cost of production, distribution expenses, staff turnover, and others. There might be a lot more variables. In order to ensure commercial success or successful strategy implementation, the firm must describe the important success elements in great detail. It is necessary to quantify these aspects in order to determine if a certain success criteria is being met.

Consistency – Knowing if the comprehensive and integrated strategy is in line with the objectives, environmental assumptions, and internal circumstances is crucial. The goals must be aligned with social responsibility, and the performance requirements must be connected to crucial success elements. Assessing the plan's ability to take full advantage of both local and global prospects, as well as its ability to keep up with technical advancements through R&D and manufacturing, is necessary to ensure that it is compatible with environmental assumptions. Coordination of procedures, rules, and the distribution of resources

Appropriateness: The suitability of a strategic decision must be assessed in light of resource capabilities, risk preferences, and time horizon.

Workability: It is important to comprehend whether the comprehensive plan is :

Feasible: in terms of the resources on hand and managerial skills. The strategy must be feasible and avoid creating problems along the road for which there is no known fix.

Stimulation – Managers and executives need to be committed to the overall strategy and have confidence that it can be implemented effectively. Before a new strategy is put into practise, all of the aforementioned qualitative requirements must be satisfied. The strategist may reevaluate the strategy, tweak it, or look into alternatives that are more appropriate for the organisation if there is not enough agreement on one or more of these concerns.

These criteria basically serve the purpose of ensuring that all aspects of the process have been adequately integrated.

Strategic Control

Control over strategy is a metaphor for shepherd control. A strategy's initial execution and the accomplishment of its planned results typically take a long period. Numerous initiatives, investments, and activities are carried out throughout that period to put the new approach into practise.

Types of Strategic Control

They must provide the basis for correcting the actions and directions of the firm in implementing its strategy as developments and changes in its environmental and internal situations take place.

The four basic types of strategic controls are:

1. Premise control
2. Implementation control
3. Strategic surveillance
4. Special alert control.

Strategic Control and Operational Control

Businesses must occasionally conduct a critical evaluation of the performance of their overall marketing objectives. The world we live in is one where goals, plans, strategies, and programmes are rapidly becoming obsolete. Therefore, every business should periodically review its internal and external strategic marketing strategy. By regularly assessing the strategy as it is being implemented, strategic control takes into consideration the shifting presumptions that govern the plan and also implements the necessary corrective action as and when necessary. Strategic management acts as an alert before disaster strikes. The practise of ensuring that particular duties are completed successfully and effectively is known as operational control. The operational control tries to assess the organization's performance. The majority of organisational control systems are of an operational type. Budgetary control, quality control, inventory control, production control, cost control, etc. are a few instances of operational control. Operational control is predetermined or is programmed. They have a self-regulating personality. They have an impersonal quality. As a way of control, techniques, instruments, and processes are employed. The operational control system pays minimal attention to external variables or adapts as necessary.

Conclusion

In conclusion, the strategic management process involves several phases and considerations that guide an organization's decision-making and actions. The criteria for evaluation play a crucial role in assessing the success and effectiveness of strategies. Both objective and subjective criteria, along with quantitative and qualitative methodologies, are utilized to establish evaluations. Quantitative criteria allow for a comparison of the firm's current performance with historical data or competitors' performance. Metrics such as earnings, dividends, stock prices, market share, and more are used to determine the extent to which strategic goals are being met. Consistency is another important criterion, ensuring that the comprehensive strategy aligns with objectives, environmental assumptions, and internal circumstances. It involves evaluating the plan's adaptability to local and global prospects, as well as its ability to keep pace with technological advancements. Additionally, appropriateness considers the suitability of strategic decisions based on resource capabilities, risk preferences, and time horizon. Workability focuses on assessing the feasibility of the comprehensive plan, ensuring that it is feasible and avoids potential roadblocks. Stimulation emphasizes the commitment and confidence of managers and executives in the strategy's effective implementation. These criteria ensure that all aspects of the strategic management process are adequately integrated. Strategic control, which includes premise control, implementation control, strategic surveillance, and special alert control, is essential for monitoring and adjusting strategies as environmental and internal situations evolve. It provides a basis for correcting actions and keeping the organization on track. Strategic control works hand in hand with operational control, which focuses on assessing performance and ensuring effective task completion. Operational control systems, such as budgetary control, quality control, and production control, are predominantly programmed and pay minimal attention to external variables. On the other hand, strategic control considers shifting assumptions and adapts as necessary. Overall, the criteria for evaluation and the strategic control process enable organizations to navigate the dynamic business environment effectively. By regularly assessing and adjusting strategies, organizations can optimize their performance, enhance their competitive position, and achieve their strategic objectives. Strategic management serves as a proactive alert system, identifying potential issues before they escalate into significant problems. By employing a comprehensive and integrated approach, organizations can stay agile and responsive to changing circumstances, ensuring long-term success.

References

Arnoldo C.Hax, Nicholas S. Majluf – 'The Strategy Concept And Process' – A Pragmatic Approach – Pearson Education Publishing Company, Second Edition, 2005.

Bartlett, C.A. and S.Ghoshal, Managing Across Borders, Boston, MA, Harvard Business Schook Press, 1989. Boston Consulting Group, Perspectives on Experience , BCG, 1972.

Buzzell, RD and BT Gale. 1987, The PIMS Principles: Linking Strategy to Performance, new York:Free Press

Cook, S. 1995Practical Benchmarking: A managers Guide to Creating Competitive Advantage. London"Kogan page

David, F R (2001). Chapter 7 'Implementing strategies: Management issues'. *Strategic Management Concepts.* 8th edition. New Jersey: Prentice Hall.

David, F R (2003). Chapter 7 'Implementing strategies: Management issues'. *Strategic Management Concepts.* 9th edition. New Jersey: Prentice Hall.

Derek F.Abell and John S.Hammond, Strategic Market Planning, Englewood Cliffs, N.J.Prentice-Hall, 1979, p.212.

Glueck, W F and L R Jauch. 1984 Business Policy and Strategic Management. New York:McGraw Hill.

Gupta, Gollakota & Srinivasan – 'Business Policy And Strategic Management – Concepts And Application' Prentice Hall Of India, 2005.

Hunger, J. D and Wheelen, T. L. (1998). Strategic Management. 6th edition. USA: Addison Wesley.

Hunger, J.D. and T.L.Wheelen, Strategic Management, (Reading, MA: Addison-Wesley, 1999), pp.143-44.

Jauch, Lawrence R. et al (2003) Business Policy and Strategic Management (First edition) Frank Bros. & Co. New Delhi.

Jeroen De Flander (2017). Strategy Implementation: Meaning and steps in Implementing Strategy.

Jeyarathmam M., Business Policy And Strategic Management, Himalaya Publishing House Pvt. Ltd., Mumbai

Johnson, Gerry & Kevan Scholes, Exploring Corporate Strategy- Text and Cases, PHI, 1996,p.225-226.

L.R.Jauch & W.F.Glueck, Business Policy and Strategic Management, Fifth Edition, McGraw Hill Intl.Edition, 1988.

Miller, A. (1998). Chapter 4 'Internal analysis'. In Strategic Management. 3rd edition. USA: McGraw-Hill, pp. 146-153.

Montanari, J. R; Morgan, C. P and Bracker, J. (1990). 'The choice perspective of strategic management'. In Strategic Management: A Choice Approach. Chicago: The Dryden Press.

Peters, T.J. & Waterman, R.H.Jr.(19820, In Search of Excellence, New York, Harper & Row.

Rao Subba P., Business Policy And Strategic Management: Text And Cases, Himalaya Publishing House Pvt. Ltd., Mumbai, 2013

Slatter, S. Corporate Recovery: Successful Turnaround Strategies and Their Implementation, (New York: Penguin, 1984)

Thomas, J.G. Strategic Management – Practices and Cases, (New York: Harper and Row, 1988), p.215.

Thomson, A.A.Jr. and A.J.Strickland, Strategic Management- Concepts and Cases, 3rd Edn, Plano, Texas; Business Publications, 1984), p.90.

Walter E.Ketchell III, Oh Where Oh Where Has My Little Dog Gone? Or My Cash Cow? Or My Star?, Fortune, Nov.2, 1981, pp.148-52.

Wheelen Thomas, Hunger J. David And Rangaranjan, Strategic Management And Business Policy, Pearson Education, 2007

17

Farm Accountancy: Terms, Concepts and Different Types of Accounts

Deepak Bhardwaj[1], Subodh Agarwal[2] and Aarti Bajwan[3]

[1]Department of Agri-Business Management, CCS HAU, Hisar, Haryana, India
[2]Assistant Professor, Department of Agri-Business Management, CCS HAU, Hisar, Haryana, India
[3]Department of Agricultural Economics, CCS HAU, Hisar, Haryana, India

Abstract

To assess farm performance, managerial control, budgeting, gains and losses sustained on the farm, records and accounting are essential. With some conditions, income from agriculture is tax-free in India. However, the evaluating officer wants to see the record that is required for farmer to produce the accounts. Therefore it is advisable to the farmers to maintain the farm accounting books. Farms are managed and run by farming families in India. The proper financial accounting is not maintained on farms. Proper maintenance of accounts is useful for owners, government and research agencies. Lack of proper farm management and accounting system are the main reason for low performance of farms in India.

Keywords: Farm accounts, performance, record, management.

Introduction

It is commonly believed that Indian farmer is a better producer than a good business man. He knows how to produce but he does not know how to keep accounts of the farm. It is therefore very necessary to have sufficient knowledge about farm accountancy in order to handle the farm business efficiently.

A crucial component of effective cattle business management is record keeping. Farmers must rely on their memory when making decisions about their farming practises because they lack written records. The dairy business keeps a number of useful records, including those of production and financial activities. According to Mahatma Gandhi, India's villages are where its spirit is found. The issues with water, power, and labour in India are making farming uncertain. As a result, India's farmers remain underprivileged. However, there are always exceptions, and in India, some farmers are also extremely wealthy. Therefore, forcing them to create farm accounting is really required. and turn the farming into industry. . The export of food grains generates foreign currency in addition to providing for people's daily needs. The farming industry also contributes to the high labour employment. The keeping of agricultural farm accounts is crucial. Keeping correct farm records and building and maintaining a tried-and-true record-keeping system are essential for farmers who also need to be financial managers. To keep track of all their business activities, farmers want an accurate bookkeeping, financial planning, and farm records system. With some conditions, income from agriculture is tax-free in India. However, the evaluating officer wants to see the record that is required for farmer to produce the accounts. Therefore it is advisable to the farmers to maintain the farm accounting books. No doubt, an average farmer in our country is illiterate. Illiteracy is not an excuse to avoid accounting.

Even though India's economy has experienced fast industrial development, it still relies heavily on agriculture, and more than 75% of the population is either directly or indirectly involved in it. More than 700,000 villages in the nation are involved in activities based on agriculture. Agriculture contributes a significant portion of the country's national income. Despite the fact that agriculture is the most important sector of our economy, no complete, in-depth, or systematic accounting, costing, or management procedures have been established specifically for this sector. Therefore it is advisable to the farmers to maintain the farm accounting books.

Farm Accountancy

Accountancy is a tool that farmers can use to closely monitor their farm business. Additionally, it aids in the management of any side businesses that farmers have started.

The practise of managing and keeping track of the financial transactions and records associated with farming operations is referred to as "farm accounting." It involves recording, classifying, summarising, analysing, and interpreting financial information to make informed decisions regarding the farm's financial health and profitability.

Farm accountancy refers to the specialized branch of accounting that deals with the financial management and record-keeping practices specific to agricultural businesses, such as farms and ranches. It involves the application of accounting principles and techniques to track, analyse, and report financial information related to farming activities. The main objective of farm accounting is to give farmers accurate and trustworthy financial data so they can make wise business decisions.

The term "Farm Bookkeeping" is also used to refer to farm accounting or accountancy. Farm accounting is an application of the accounting principles to the business of farming.

Farm bookkeeping is known as a system of records written to furnish a history of the business transactions, with special reference to its financial side. The Farm Accountancy or Farm Book Keeping is a matter of Farm Business Analysis.

Key aspects of farm accountancy

i. **Book keeping:** Maintaining thorough records of all financial transactions, including receipts and payments for all purchases, sales, loans, and other transactions. You have two options for doing this: manually or with specialised accounting software.

ii. **Financial Statements:** Preparing financial statements like balance sheets, cash flow statements, and income statements (profit and loss statements). These financial statements offer a summary of the farm's assets, liabilities, and cash flow, which can be used to identify problem areas and make strategic decisions.

iii. **Tax Planning:** Understanding and complying with tax regulations specific to agriculture. This includes taking advantage of tax deductions, exemptions, and credits available to farmers.

iv. **Risk Management:** Implementing strategies to mitigate financial risks associated with farming, such as market fluctuations, weather events, and commodity price volatility. This may involve insurance coverage, diversification, and hedging strategies.

v. **Income and Expense Tracking:** Farm accountancy involves recording all income and expenses associated with farming operations. This includes tracking revenue from crop sales, livestock sales, government subsidies, and other sources, as well as expenses such as seed, fertilizer, equipment maintenance, labour costs, and utilities.

vi. **Inventory Management:** Farm accountancy encompasses the management of inventories related to crops, livestock, and agricultural inputs. Farmers must monitor the value of their inventories, take into account variations in quantity and quality, and compute cost of goods sold.

vii. **Depreciation and Asset Valuation:** Farms often possess a range of assets such as land, buildings, machinery, and vehicles. Accurately estimating these assets' current values and calculating their rate of depreciation over time are key components of farm accounting.

viii. **Budgeting and Financial Planning:** Farm accounting includes creating and maintaining financial plans and budgets for the farm business. Forecasting income and expenses, establishing financial objectives, and comparing actual performance to the budget are all part of this process.

ix. **Cash Flow Management:** Farm accountancy focuses on controlling the farm's cash flow and making sure there is enough money to cover ongoing costs and capital requirements. It entails monitoring cash inflows and outflows, controlling credit and loans, and managing working capital effectively.

Farm accountancy is crucial for farmers to have a clear understanding of their financial position, make informed decisions, secure financing, and comply with legal and tax requirements. To ensure accurate financial management and maximise profitability, many farmers enlist the help of certified public accountants or agricultural consultants with knowledge of farm accounting.

Important Terms and Concepts Related to Farm Accountancy

1. **Account:** An account is a detailed outline of the transactions that are carried out by a specific business in respect of a particular person or a firm or their representatives or objects.

For example, when a business carries out transactions with customers and suppliers, both suppliers, as well as customers, are termed separate accounts.

Similarly, businesses sometimes purchase tangible items like land, machinery, plant, building, etc., and each of the tangibles is treated as an individual account though such types of accounts are related to things.

2. **Accounting:** Accounting is the process of recording, classifying, summarizing, analysing and reporting financial transactions of a business or an individual.

It involves the systematic and organized recording of financial data to provide accurate and useful financial information for decision-making purposes. Accounting follows specific principles, standards, and frameworks to ensure consistency and comparability in financial reporting.

3. **Accountancy:** The term "accountancy" is more general and refers to the profession, principles, and methods of accounting. It refers to the work performed by accountants, including the preparation, analysis, and interpretation of financial statements, tax planning, auditing, and financial advisory services.

4. **Single Entry System:** The single-entry system is a simple book keeping method that records only one side of the accounting entry for each transaction. It is commonly used by small businesses and individuals who maintain a simple record of cash inflows and outflows. In the single-entry system, transactions are recorded in a cash book, and the financial statements are prepared based on the available information. However, this system lacks the robustness and accuracy of the double entry system. This system is incomplete and inaccurate.

5. **Double Entry System:** The double entry system is a comprehensive method of bookkeeping that records two sides (debit and credit) for each financial transaction. It follows the fundamental accounting equation, which states that assets are equal to liabilities plus equity. Every transaction affects at least two accounts, with one account debited and another credited. This system provides a more accurate and complete picture of a company's financial position and performance and is widely used in businesses of all sizes.

The double-entry system of bookkeeping has several advantages that make it widely used and highly effective. Here are some of the main advantages:

1. **Accuracy and Error Detection:** The double-entry system is made to make sure that financial transactions are recorded accurately. Each transaction is entered twice, once as a credit and once as a debit. This balance of credits and debits helps in the error detection process because any discrepancy in the totals will point to an error in the recording.

2. **Financial Statement Preparation:** The double-entry system provides the necessary framework for preparing financial statements. It enables the creation of accurate balance sheets, income statements, and cash flow statements by recording transactions in the appropriate accounts. This enables companies to evaluate their financial situation and make wise decisions.

3. **Enhanced Financial Control:** The system allows for better financial control by providing a comprehensive view of the financial position of a business. Through double-entry bookkeeping, businesses can track income, expenses, assets, liabilities, and equity. This information is crucial for monitoring cash flows, analysing profitability, and managing resources effectively.

4. **Facilitates Audit and Tax Compliance:** The auditing process is made easier by double-entry bookkeeping's systematic approach. Businesses can easily provide proof of their financial transactions by keeping accurate records and supporting documentation, which makes audits more effective. The system also helps with tax compliance by making sure that earnings, expenses, and tax obligations are accurately recorded.

5. **Easy Identification of Financial Trends:** Double-entry bookkeeping makes it possible to recognise trends and patterns, which facilitates insightful financial analysis. Businesses can evaluate their financial performance, pinpoint areas for improvement, and make well-informed decisions for the future by comparing data from various accounts and time periods.

6. **Facilitates Business Planning and Decision-Making:** Businesses can plan and decide wisely when accurate and current financial information is available. Through insights into cash flows, profitability, and financial ratios, the double-entry system enables businesses to evaluate their financial viability and develop growth or expansion plans.

7. **Reliable Basis for Financial Transactions:** The double-entry system offers a reliable basis for recording and monitoring financial transactions. Its standardised methodology ensures consistency and uniformity, facilitating communication and collaboration between accountants, bookkeepers, and financial professionals.

In general, there are many benefits to using the double-entry system of bookkeeping, including accuracy, error detection, financial control, audit compliance, financial analysis, and well-informed decision-making. It is a widely used and reliable approach that offers a strong framework for keeping financial records and successfully managing business finances.

An accounting system is a set of processes, procedures, and software tools used by businesses to record, analyse, and report financial transactions. In addition to assisting businesses in maintaining accurate financial records for decision-making, tax compliance, and financial reporting purposes, it offers a structured way to track income, expenses, assets, liabilities, and equity.

Here are some key components and features typically found in an accounting system:

i. **Chart of Accounts:** A list of all the accounts used to classify and keep track of financial activities. Assets, liabilities, equity, revenue, and costs are just a few of the areas it covers.

ii. **Accounts Payable:** These accounts keeps track of and administers the money owing to creditors such as suppliers, vendors, and other parties.

iii. **Accounts Receivable:** These accounts keeps track of and handles the money that clients owe for the sale of products or services.

iv. **Financial Statements:** Financial Statements are the documents produced by the accounting software, including the cash flow statement, income statement, and balance sheet. These statements provide an overview of the financial position and performance of the business.

v. **Bank Reconciliation:** Bank Reconciliation is the process of matching and comparing the balances in the accounting system with the bank statements to make sure they are in harmony.

vi. **Budgeting and Forecasting:** Budgeting and forecasting are the features that allow businesses to set financial goals, create budgets, and project future financial performance.

vii. **Reporting and Analysis:** Tools and skills that let organizations' to create different financial reports and analyse financial data to learn more about the performance and health of the company's finances.

viii. **Audit Trail:** A record of all accounting activities and changes made in the system, providing transparency and accountability.

ix. The accounting cycle, which is used in bookkeeping and financial reporting, includes the following crucial elements: Journal, Ledger, Trial Balance, and Final Accounts. They are used to create financial statements to record and to summarize financial activities.

   1. **Journal:** A journal is a book or electronic record where all financial transactions of a business are initially recorded in chronological order. It serves

as a primary record of entries and contains details such as the date, description of the transaction, accounts affected, and the corresponding amounts. The process of recording transactions in a journal is called journalizing.

2. **Ledger:** A ledger is a group of accounts that includes separate entries for every kind of asset, obligation, equity, income, and cost. All the transactions listed in the journal are kept in one place at this one location. The ledger summarizes and organizes transactions to give a complete picture of a company's financial activities. It is used to create financial statements and gives a summary of financial transactions. The process of transferring information from the journal to the ledger is called posting.

3. **Trial Balance:** A trial balance is a statement that lists the balances of all the accounts in the ledger at a specific point in time. It ensures that the total debits and credits recorded in the ledger are equal, thus serving as a preliminary check on the accuracy of the recording and posting process. The trial balance helps identify errors or discrepancies before the preparation of financial statements.

4. **Final Accounts:** Final accounts are the financial statements prepared at the end of an accounting period to provide a summary of a business's financial performance and position. The two main types of final accounts are the Trading Account and the Profit and Loss Account (also known as the Income Statement), which present the revenues, expenses, and profitability of the business. The final accounts also include the Balance Sheet, which shows the assets, liabilities, and owner's equity of the business.

In summary, the accounting cycle typically involves recording transactions in a journal, transferring them to the ledger, preparing a trial balance to verify accuracy, and finally generating the final accounts to present the financial results and position of a business. These steps ensure the systematic and accurate recording, summarizing, and reporting of financial information.

5. **Subsidiary books,** also known as subsidiary journals, are specialized accounting books that help in the systematic recording of transactions related to specific categories. One common type of subsidiary book is the cash book, which is used to record all cash transactions of a business.

6. **The cash book** is an essential subsidiary book that tracks cash receipts and cash payments. It serves as a primary record of all cash inflows and outflows, allowing for easy monitoring and reconciliation of cash balances. The cash

book usually has two sides: the debit side, which records cash receipts, and the credit side, which records cash payments.

Here are the key features and benefits of maintaining a cash book:

1. **Efficient cash management:** The cash book helps in efficiently managing cash by providing a clear picture of cash inflows and outflows. It enables businesses to monitor their cash position, identify surplus or deficit, and make informed decisions accordingly.
2. **Quick and easy recording:** Transactions involving cash are recorded directly in the cash book, eliminating the need for separate entries in the general ledger for each transaction. This saves time and effort in maintaining the accounts.
3. **Internal control:** The cash book acts as an internal control mechanism, as all cash transactions are recorded in a single book. This facilitates easy auditing, reconciliation, and identification of any discrepancies or errors.
4. **Bank reconciliation:** The cash book serves as a basis for bank reconciliation. By comparing the cash book balances with the bank statement, businesses can identify any discrepancies and rectify them promptly.
5. **Analysis and reporting:** The cash book provides valuable data for analysis and reporting purposes. It allows businesses to track cash inflows and outflows by various categories, such as revenue, expenses, or specific cash accounts, enabling effective financial analysis and decision-making.

It's important to note that while the cash book is a type of subsidiary book, there are other subsidiary books as well, such as the sales book, purchases book, journal proper, and ledger subsidiary books (for recording transactions related to individual accounts). These subsidiary books help in maintaining accurate and organized accounting records for specific types of transactions.

Accounting systems can be manual, where transactions are recorded in physical books, or computerized, where accounting software is used to automate and streamline the process. Common accounting software packages include QuickBooks, Xero, Sage, and SAP.

The choice of an accounting system depends on the size and complexity of the business, industry-specific requirements, budget, and other factors. It is essential for businesses to have an effective accounting system in place to ensure accurate financial reporting, compliance with regulations, and informed decision-making.

Types of Accounts

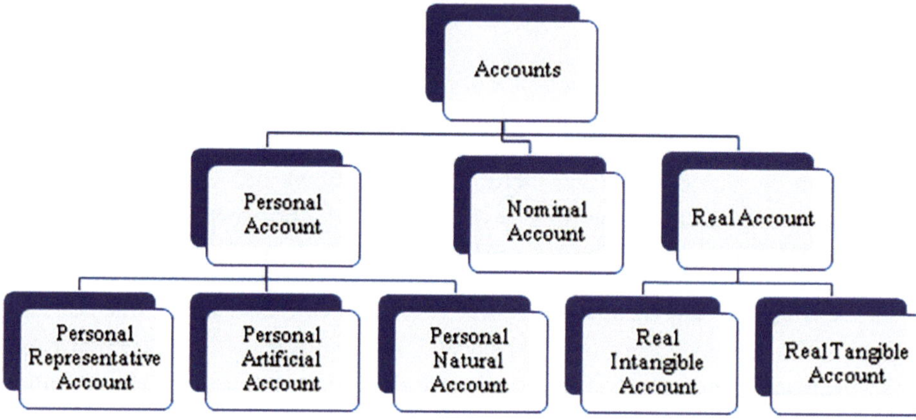

A Real Account can also be called a general ledger account that relates to assets and liabilities other than people accounts. These accounts are accounts that don't need to be closed at the end of the financial year because they are carried forward to the next year. A simple example of a real account is a bank account.

A personal account can be called a general ledger account that is connected to all persons or people like individuals, different firms, or associations. A creditor account is an example of a personal account. A nominal account is a general ledger account that keeps a record of all income and expenditure, gains and losses. A simple example of a nominal account is an interest account.

1. Personal Account

As stated earlier, personal accounts are those accounts that are related to an individual, a company, a firm or a group of associations, etc. These persons might incorporate natural persons, artificial persons, or representative persons, as the case may be.

For example – Manoj and Saroj Trading Co., Charitable Trusts, ABC Bank Ltd, X company Ltd., etc.

There are some accounts that might come under the category of personal representative accounts.

For instance – When we speak of salary, it means how much amount is payable to each of the employees. But all salary accounts are clubbed collectively under an account called 'salary payable A/c'.

The rule for this Account:

Debit The Receiver, Credit The Giver.

Type of Personal Accounts

a. Natural Persons

These types of accounts are related to individuals or natural persons like Rajkumar's A/c, Anil's A/c, Ritul's A/c, etc.

b. Artificial Accounts

These accounts are related to various companies and institutions like Roy Brothers Pvt Ltd A/c, Lion's Club A/c, etc. Thus, such types of institutions and companies are those entities that are there in the eyes of law.

c. Representative Accounts

Accounts that represent a specific purpose of work are called representative accounts. For instance, Outstanding Wages A/c, Outstanding Interest A/c, Prepaid Expense A/c, etc.

The golden rule that is related to the personal account is:

1. The receiver is debited
2. The giver is credited.

Illustration

Siddharth bought some machinery from M/s Surana & Sons worth Rs 10,00,000/- on credit. So we see that this particular transaction affects two accounts: first, the Personal Account of M/s Surana & Sons and the Machinery Account. Thus, this transaction implies that Siddharth has purchased the Machinery from M/s Surana & Sons for his business. The Golden Rule of Personal Account states, "Debit the Receiver, Credit the Giver".

As in this transaction, M/s Surana & Sons is the Giver, his Personal Account shall be credited with Rs 10,00,000. On the other hand Machinery, A/c shall be debited with the same sim of money. Therefore, this particular transaction is being recorded in the following manner in the respective accounts:

Machinery Account			
Particulars (Dr)	Amount (Rs)	Particulars (Cr)	Amount (in Rs)
To M/s Surana	10,00,000		
M/s Surana			
Particulars (Dr)	Amount (in Rs)	Particulars (Cr)	Amount (in Rs)
		By Machinery	10,00,000

2. Real Account

Real Accounts are those accounts that relate to assets, properties, or possessions. These related properties might exist in physical or non-physical forms. This gives rise to the need for the creation of two types of real accounts:

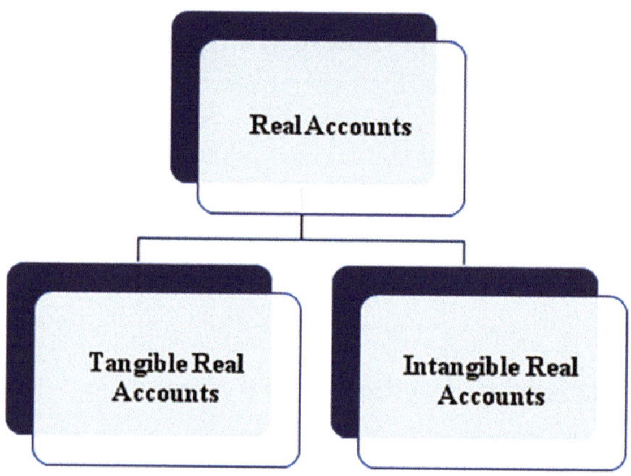

a. Tangible Real Accounts

The term tangible real accounts suggest those accounts that are physical in nature. In other words, these assets are visible to the eyes. Such assets can be touched, seen, or felt. For instance, Building A/c, Vehicle A/c, Machinery A/c, etc.

b. Intangible Real Accounts

This type of account suggests those accounts that relate to assets or possessions that are non-physical in nature. In other words, these assets cannot be seen, felt, or touched but can be measured in terms of some amount of money. One can say that some value is attached to these types of assets.

For example, goodwill, patent, trademarks, copyrights, etc.

Golden Rule relating Real account:

Debit what comes in, Credit what goes out.

Illustration

Siddharth bought a vehicle for the purpose of his business which is worth Rs 5,00,000 in cash. So, this type of transaction involves two real accounts: A vehicle Account and Cash Account.

Hence, buying a vehicle worth Rs 5,00,000 in cash implies that a vehicle is being added to the business. At the same time, cash is going out of the business. Therefore, the golden rule of real accounts states, "Debit What Comes in, Credit What Goes Out".

Both cash and vehicle are real accounts, hence, Rs 5,00,000 will be debited to vehicle A/c. But the same sum will be credited to cash A/c.

So, the following table shows the transaction thus recorded:

Vehicle Account			
Particulars (Dr)	Amount (in Rs)	Particulars (Cr)	Amount (in Rs)
To Cash	5,00,000		
Cash Account			
Particulars (Dr)	Amount (in Rs)	Particulars (Cr)	Amount (in Rs)
		By Vehicle	5,00,000

3. Nominal Account

Nominal accounts are those types of accounts that are related to any form of income or expenditure, gain or loss. For example, Rent A/c, Salary A/c, Wages A/c, etc.

The golden rule for such accounts:

Debit all the expenses and losses, Credit all the incomes and gains

For Example – Whenever any salary is given to employees of the business entity salary A/c is debited or whenever any other expenses are incurred it is debited. On the contrary, when the business gets any discount, interest, etc these are credited whenever received by the business entity.

In addition, there are some other types of accounts in accounting which are as follows:

1. **Revenue Accounts:** These accounts track the income generated by the farm, such as crop sales, livestock sales, and other sources of revenue.
2. **Expense Accounts:** These accounts record various expenses incurred in farming operations, including seed, fertilizer, feed, labour, utilities, equipment maintenance, and repairs.
3. **Asset Accounts:** These accounts represent the value of assets owned by the farm, such as land, buildings, machinery, vehicles, and livestock.
4. **Liability Accounts:** These accounts track any debts or obligations of the farm, such as loans, mortgages, or outstanding payments to suppliers or creditors.
5. **Equity Accounts:** These accounts represent the owner's equity or net worth of the farm. Equity increases with profits and contributions from the owner and decreases with losses or withdrawals.
6. **Cash Accounts:** These accounts record the cash inflows and outflows of the farm, including sales receipts, payments to suppliers, and other cash transactions.

These are some of the key concepts and types of accounts involved in farm accountancy. Effective farm accountancy practices enable farmers to assess the financial performance of their operations, make informed decisions, and maintain the financial health of their businesses.

All transactions of a business entity should be recorded in account books. To record these transactions the business entity should pass journal entries which will be then entered into ledgers. The journal entries are passed according to the Golden Rules of Accounting. In order to apply these rules, the type of account is to be ascertained first and then these rules are applied thereafter:

» Debit is something that comes in, credit is something that goes out

» The receiver is to be debited, the giver is to credit

» All expenses are to be debited and all income is to be credited

So in this way, the foundation of accounting is laid. Above mentioned rules are known as the Golden Rules of accounting. These rules are just like the letters of the

English alphabet. For instance, one can write in English only if he knows how to write the English alphabet. Similarly for accounting, if one does not have knowledge of the above-mentioned golden rules, one will not be able to pass journal entries and therefore will not be able to account accurately for the transactions.

Conclusion

The farms records and accounts are important tools in farm management and are in fact, the sine- qua non (an essential condition; a thing that is absolutely necessary) of effective farm management, farmers all over the world at whatever level should be encouraged to keep records of all the activities carried out on the farm from the beginning to the end of each farming season and be accountable for effective present and future performances. In Indian villagers literacy rate is very low. As we know 48% Indians are depending on agriculture and most of the villagers are employed in this sector. A big farmer's wealth is as good as small scale industries. But he has not maintaining his accounting system. The use of well-developed farm accounting by the farmers can change the present scenario of agriculture.

References

Sharma, J. K., and Dubey, P. K. (2019). A Need of Farm Records and Accounting in Agriculture Sector. *International Journal of Innovative Studies in Sociology and Humanities*, **4**(5), 155-158.

Raju, V. T., Rao, D. V. S., (1990). Economics of Farm Production and Management. Oxford and IBH Publishing Company Private Limited. ISBN-13: 978-81-204-0423-6.

Reddy, S. S., Ram, P. R., Sastry, T. N., and Devi, I. B. (2018). Agricultural Economics. Oxford and IBH Publishing Company Private Limited. ISBN: 978-81-204-1786-1.